Transmission Line Transformers

Transmission Line Transformers is a member of the Noble Publishing Classic Series. Titles selected for Classic Series Membership are important works which have stood the test of time and remain significant today. While many publishers lose interest in old titles, Noble Publishing is proud to recognize classic works and to revive their contribution.

NOBLE PUBLISHING PRODUCTS

CLASSIC SERIES TITLES

Electronic Applications of the Smith Chart
    Phillip H. Smith

Microwave Field-Effect Transistors
    Raymond S. Pengelly

Microwave Transmission-Line Impedance Data
    M. A. R. Gunston

Transmission Line Transformers
    Jerry Sevick

OTHER TITLES

HF Filter Design and Computer Simulation
    Randall W. Rhea

Oscillator Design and Computer Simulation
    Randall W. Rhea

Transceiver System Design for Digital Communications
    Scott R. Bullock

SOFTWARE

*win*Smith
*win*LINE
EEpal

VIDEOS

Introduction to the Smith Chart
Filters and Matching Networks
Oscillator Design Principles
RF/Microwave Transistor Amplifier Design
Microwave Filters, Couplers and Matching Networks
RF Circuit Fundamentals I
RF Circuit Fundamentals II
Microwave Transmission Lines and their Physical Realizations

Transmission Line Transformers

by
Jerry Sevick, W2FMI

Noble Publishing
Atlanta

Sevick, Jerry
    Transmission Line Transformers/Jerry Sevick
    3rd ed.
    Includes bibliographical references.
    ISBN 1-884932-66-5
    1. Baluns and transformers-design and construction
    2. Transmission line transformers
    3. High frequency transformers
I. Title

To order contact:
    Noble Publishing
    2245 Dillard Street
    Tucker, Georgia 30084 USA
    (770)908-2320
    (770)939-0157

Discounts are available when ordered in bulk quantities.

N⊕BLE

Printed and bound in the United States of the America
10 9 8 7 6 5 4 3 2 1

International Standard Book Number 1-884932-66-5

Preface

Although I was gratified at the response I received from readers of the first edition of *Transmission Line Transformers*, it became very apparent that more specific information was needed on practical designs, construction techniques and sources of materials. This request was, by far, the dominant one in the many Feedback forms that were returned to the publisher. A few requests were concerned with transformers for antenna tuners, hybrids, and the VHF and UHF bands. From the professionals, the request was for more information on the "third wire" in the 1:1 balun.

The feedback from the first edition (deeply appreciated by the author) led to this second edition, which attempts to respond to these many requests. Although the design goal was to supply a great variety of transformers for matching 50-Ω cable to antennas in the 1.5- to 30-MHz range (which has been the interest and experience of the author), many of them should perform well in other areas. The new analytical information and the different viewpoint on baluns presented in this second edition should also be of help.

In the process of designing some one hundred new transformers (hopefully, practical ones) for this second edition, many interesting and unexpected designs emerged. Specifically, these new designs resulted from the use of Guanella's approach to these wideband transformers, which was published in 1944 (ref 1).[1] His technique was to connect transmission lines in a parallel-series arrangement such that in-phase voltages were summed at the high-impedance side. The transmission lines were appropriately coiled to prevent the unwanted (transformer) currents. His simple and important statement, "a frequency independent transformation," which appeared in his paper, had been overlooked by this author (and probably by others), as is evidenced by the scarcity of his designs in the first edition. On the other hand, Ruthroff, who published in 1959 (ref 9), obtained a 1:4 transformation ratio by summing a direct voltage with a delayed voltage which traversed a single transmission line. By connecting the transmission line such that a negative potential gradient existed, the transformer became a balun;

---

[1]Each reference can be found in Chapter 15.

with a positive gradient, it became an unun (unbalanced-to-unbalanced transformer). Even with transmission lines having optimized characteristic impedances, his transformers had a built-in high-frequency cut-off.

Since their two approaches to these wideband transformers are distinctly different, the author has taken the liberty of using their names in describing which design is used. In this way, they are not only given credit for their important contributions, but it has also been found to be an apt description. Each approach has its specific application in the field of transmission line transformers. This edition attempts to supply designs which utilize the best approach for the objectives on hand.

This second edition also attempts to dispel some of the misconceptions regarding transmission line transformers. These misconceptions arise from perceiving these devices as conventional transformers. Therefore, most of the concerns are related to the properties of the core such as size, loss, saturation, IMD (intermodulation distortion), frequency response and so on. Other conventional transformer properties usually considered are magnetizing current, turns ratio and primary and secondary windings. As is shown in this edition, with properly designed transmission line transformers, all these concerns are really irrelevant.

I wish to acknowledge the help of the many companies who supplied, generously and promptly, materials much needed for this second edition. They are listed in Chapter 11, along with their addresses and telephone numbers (where available).

I again thank my following colleagues from AT&T Bell Laboratories: M. D. Fagen for reviewing my second edition manuscript in great detail and for his continuous interest and very helpful suggestions, and C. L. Ruthroff for his technical support and important discussions on baluns. I also owe a debt of gratitude to The American Radio Relay League Inc for their fine support in connection with this second edition. I take pleasure in especially thanking Gerald L. Hall, ARRL Associate Technical Editor, and Jeffrey S. Kilgore, ARRL Assistant Technical Editor, for their excellent technical help and cooperation in bringing this second edition to press.

Finally, this second edition is dedicated to my wife, Connie. I owe her much for her encouragement and sympathetic understanding through the preparation of this second edition. She has earned this dedication.

Jerry Sevick, W2FMI
Basking Ridge, New Jersey
March 1990

Contents

Chapter 1

Analysis

Sec 1.1   Introduction

T here are two basic methods for constructing broadband, imped-
ance-matching transformers. One employs the conventional
transformer that transmits the energy to the output circuit by
flux linkages; the other uses the transmission line transformer to trans-
mit the energy by a transverse transmission line mode. With techniques
exploiting high magnetic efficiency, conventional transformers have
been constructed to perform over wide bandwidths. Losses on the order
of one decibel can exist over a range from a few kilohertz to over
200 MHz. Throughout a considerable portion of this band, the losses
are only 0.2 dB.

On the other hand, transmission line transformers exhibit far
wider bandwidths and much greater efficiencies. The stray inductances
and interwinding capacitances are generally absorbed into the characteristic impedance of the transmission line. As such, they form no
resonances that could seriously limit the high-frequency response. Here
the response is limited by the deviation of the characteristic impedance
from the optimum value; the parasitics not absorbed into the characteristic impedance of the transmission line; and, in some transformer
configurations, the length of the transmission line.

With transmission lines, the flux is effectively canceled out in the
core and extremely high efficiencies are possible over large portions of
the passband—losses of only 0.02 to 0.04 dB with certain core materials.

Therefore, the power ratings of transmission line transformers are
determined more by the ability of the transmission lines to handle the
voltages and currents than by the size and conventional properties of
the core.

The earliest presentation on transmission line transformers was
by Guanella in 1944 (ref 1).[1] He proposed the concept of coiling

---

[1]Each reference in this chapter can be found in Chapter 15.

transmission lines to form a choke that would reduce the undesired mode in balanced-to-unbalanced matching applications. Before this time, this type of device, known as a *balun*, was constructed from quarter- or half-wavelength transmission lines and, as such, had very narrow bandwidths. By combining coiled transmission lines in parallel-series arrangements, he was able to demonstrate broadband baluns with ratios of $1:n^2$, where n is the number of transmission lines. Other writers followed with further analyses and applications of the balun transformer introduced by Guanella (refs 2-8). In 1959, Ruthroff published another significant work on this subject (ref 9). By connecting a single transmission line such that a negative or a positive potential gradient existed along the length of the line, he was able to demonstrate a broadband 1:4 balun or *unun* (unbalanced-to-unbalanced) transformer. He also introduced in his paper the hybrid transformer. Many extensions and applications of his work were published and are included in the reference list (refs 10-28).

In a general comparison, it can be said that the transmission line transformer enjoys the advantage of higher efficiency, greater bandwidth and simpler construction. The conventional transformer, however, remains capable of dc isolation. The purpose of this chapter is twofold: to review Guanella's and Ruthroff's approaches to the analysis and understanding of these new wide-band transformers, and to present additional material to form a basis for the chapters that follow.

Sec 1.2   The Basic Building Block

The single bifilar winding, shown in Fig 1-1, is the basic building block for the understanding and design of all transmission line transformers. Higher orders of windings (trifilar, quadrifilar and so on) also perform in a similar transmission-line fashion and will be discussed in Chapter 7, which treats the subject of fractional-ratio transformers.

The circuit of Fig 1-1 can perform four different functions depending upon how the output load, R_L, is grounded. The functions are: (A) a phase-inverter when a ground is connected to terminal 4, (B) a balun when the ground is at terminal 5 or left off entirely (a floating load), (C) a simple delay line when a ground is at terminal 2, and (D) a "boot-strap" when $+V_1$ is connected to terminal 2. The operation of these four functions can be explained by simple transmission line theory and the choking reactance of the transmission lines. This choking reactance, which isolates the input from the output, is usually obtained

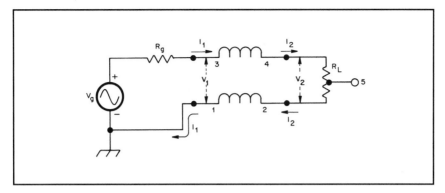

Fig 1-1—The basic building block.

by coiling the transmission line around a ferrite core or by threading the line through ferrite beads. The objectives, in practically all cases, are to have the characteristic impedance Z_0, of the transmission line equal to the value of the load, R_L (which is called the optimum characteristic impedance) and to have the choking reactance of the transmission line much greater than R_L (and hence Z_0). Meeting these objectives results in a "flat" line and hence maximum high frequency response and maximum efficiency since conventional transformer currents are suppressed. In the final analysis, the maximum high-frequency response is determined by the parasitics not absorbed into the characteristic impedance of the line, and the efficiency by the properties of the ferrites when used in *transmission line transformer* applications.

A further understanding of transmission line transformers can be gained by noting the longitudinal potential gradients that exist with the circuits of the following four:

A) *Phase Inverter*

By connecting a ground to terminal 4, a negative potential gradient of $-V_1$ is established from terminal 3 to 4. The gradient from terminal 1 to 2 is $-V_2$. For a matched load, $V_1 = V_2$. If the reactance of the windings (or a straight transmission line loaded with beads) is much greater than R_L, then only transmission line currents flow and terminal 2 is at a $-V_2$ potential. When the reactance is insufficient, a shunting, conventional current will also flow from terminal 3 to 4, resulting in a drop in the input impedance and the presence of flux in the core. As the frequency is decreased, the input impedance approaches zero.

B) *Balun*

By connecting a ground to terminal 5, a negative potential gradient $-(V_1 - V_2 / 2)$ is established from terminal 3 to 4 and $-V_2 / 2$ from terminal 1 to 2. With a matched load, $V_1 = V_2$ and the output is balanced to ground. When the reactance fails to be much greater than R_L, conventional transformer current will flow and eventually, with decreasing frequency, the input impedance approaches $R_L / 2$. When the load is "floating," the currents in the two windings are always equal and opposite. At very low frequencies, where the reactance of the windings fails to be much greater than R_L, the isolation of the load is inadequate to prevent conventional transformer current (which could be an antenna current) when the load is elevated in potential. This bifilar balun, which was first proposed by Guanella (ref 1), is completely adequate for most 1:1 balun applications when the reactance of the windings (or beaded straight transmission lines) is much greater than R_L.

C) *Delay Line*

By connecting the ground to terminal 2, the potential gradient across the bottom winding is zero. With a matched load, the gradient across the top winding is also zero. Under these conditions, the longitudinal reactance of the windings plays no role. The transmission line simply acts as a delay line and does not require winding about a core or the use of ferrite beads. This delay function plays a most important role in obtaining the highest frequency response in unbalanced-to-unbalanced transformers.

D) *Boot-Strap*

Probably the most unlikely circuit schematic with the basic building block is the one where $+V_1$ is also connected to terminal 2 (that is, terminal 3 is connected to terminal 2). By this type of connection, a positive potential gradient of V_1 is established across the bottom winding and of $+V_2$ across the top winding. When the bottom of R_L is connected to ground, instead of to terminal 2, a voltage of $(V_1 + V_2)$ exists across its terminals. This "boot-strap" connection, in which the transmission line shares a part of the load, is the way Ruthroff (ref 9) obtained his 1:4 unbalanced-to-unbalanced transformer.

Sec 1.3   The Guanella Analysis

Guanella's investigation (ref 1) was directed toward developing a broadband transformer for matching the balanced output of a 100-W, push-pull, vacuum-tube amplifier to the unbalanced load of a coaxial cable. The objective was to match a balanced impedance of 960 Ω to an unbalanced impedance of 60 Ω (16:1) from 100 MHz to 200 MHz. His experimental data, with a 53-Ω resistor as a load, showed a deviation of less than 10% from the theoretical value over this frequency range.

Guanella accomplished this by incorporating four 240-Ω transmission lines in a series-parallel arrangement resulting in a high-impedance, 16:1 balun. His technique of essentially summing in-phase voltages at the high-impedance side of the transformer had been overlooked by the author (and probably by others) as is evidenced by the scarcity of information on his designs in the literature. This section presents the high-frequency analysis of his transformers and explains the advantages of his technique with transformers for high-power and high-impedance applications. Chapter 2, which covers the low-frequency analysis of his 1:4 transformer, will discuss some interesting and important configurations. Chapters 8 and 9 will present more information on Guanella's transformers with ratios greater than 1:4.

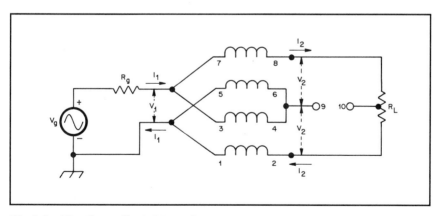

Fig 1-2—The Guanella 1:4 transformer.

Fig 1-2 is a schematic of Guanella's 1:4 transformer. The two transmission lines are in parallel on the low-impedance side and in series on the high-impedance side. With the single connection to ground, as shown in Fig 1-2, the transformer performs as a step-up balun with a floating load. With the ground connected to terminal 2 instead of the 1,5 terminal, it performs as a step-down balun with a floating load.

The high-frequency performance is determined, in large measure, by the optimization of the characteristic impedance of the transmission lines. From the symmetry of the schematic, it is quite evident that each transmission line sees one-half of the load. Therefore, for "flat" lines, and hence maximum high-frequency response, the optimum value of the characteristic impedance is $Z_0 = R_L / 2$. Without any other parasitics which are not absorbed into the characteristic impedance, this transformer (as Guanella stated in his paper) yields a "frequency independent transformation." Straight beaded lines, or having sufficient separation between bifilar windings on a core, result in near-ideal transformers.

With two transmission lines, as in Fig 1-2, the input impedance at the low side is

$$Z_{in} = \frac{Z_0}{2} \left(\frac{Z_L / 2 + j Z_0 \tan \beta l}{Z_0 + j Z_L / 2 \tan \beta l} \right) \qquad \text{(Eq 1-1)}$$

where

Z_0 = the characteristic impedance
Z_L = the load impedance
l = the length of the transmission line
$\beta = 2 \pi/\lambda$, where λ = the effective wavelength in the transmission line.

With the optimum value of $Z_0 = R_L / 2$ for a resistive load, Eq 1-1 reduces to

$$Z_{in} = R_L / 4 \qquad \text{(Eq 1-2)}$$

With more than two transmission lines, it can be shown that Eq 1-2 becomes

$$Z_{in} = R_L / n^2 \qquad \text{(Eq 1-3)}$$

where n is the number of transmission lines. Conversely, it can be seen by inspection that when looking in at the high-impedance side,

$$Z_{in} = n^2 R_L \qquad \text{(Eq 1-4)}$$

where R_L would be the low impedance on the left side in Fig 1-2.

By the placement of two ground connections, one can obtain other transformers such as ununs (unbalanced-to-unbalanced), phase-

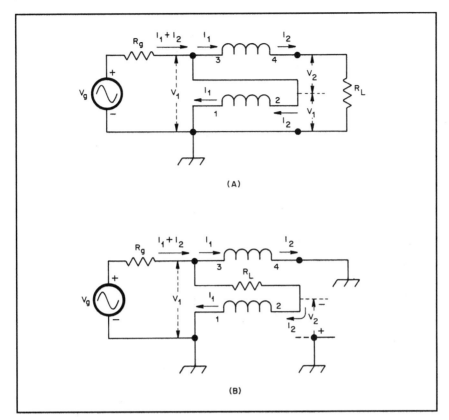

Fig 1-3—The Ruthroff 1:4 transformers: (A) unun and (B) balun.

reversals and hybrids. The low-frequency responses of these various transformers can differ greatly depending upon where the ground connections are made and the number of cores used. These will be discussed at length in Chapter 2.

Sec 1.4   The Ruthroff Analysis

Ruthroff presented, in his classical 1959 paper (ref 9), another technique for obtaining a 1:4 impedance transformation. The concept involved summing a direct voltage with a delayed voltage which traversed a single transmission line. Since his investigations involved small-signal applications, he was able to use very small, high-permeability cores and fine wires. His manganese-zinc cores ranged only from 0.175 to 0.25 inch in OD and from 1600 to 3000 in permeability. His conductors, which were twisted in order to control the characteristic impedance, were only no. 37 and 38 wires. Since the transmission lines were very short under these conditions (therefore little phase-shift between the summed voltages), he was able to demonstrate passbands essentially "flat" from 500 kHz to 100 MHz.

Fig 1-3 shows the high-frequency schematics of the two 1:4 transformers presented by Ruthroff. Fig 1-3A has the basic building block in the "boot-strap" configuration resulting in a 1:4 unun (unbalanced-to-unbalanced transformer). Fig 1-3B has the basic building block in the phase-reversal configuration resulting in a 1:4 balun. These high frequency models assume sufficient longitudinal reactances of the windings such that the outputs are completely isolated from the inputs. Unlike Guanella's model, which can practically be analyzed by inspection, Ruthroff resorted to loop and transmission line equations to solve for the power in the load and hence transducer (insertion) loss. For the unun of Fig 1-3A, they are:

$$V_g = (I_1 + I_2)R_g + V_1$$

$$I_2 R_L = V_1 + V_2$$

$$V_1 = V_2 \cos \beta l + j I_2 Z_0 \sin \beta l$$

$$I_1 = I_2 \cos \beta l + j V_2 / Z_0 \sin \beta l \qquad \text{(Eq 1-5)}$$

He also found that the maximum transfer of power occurs when $R_L = 4R_g$ and that the optimum value of the characteristic impedance is $Z_0 = 2R_g$. Fig 1-4 shows the loss as a function of the normalized line length and for various values of the characteristic impedance, Z_0. Even with the optimum value of the characteristic impedance, the loss is found to be 1 dB when the line is a quarter-wavelength and infinite when it is a half-wavelength. Fig 1-4 shows the value of keeping the transmission line as short as possible with Ruthroff's 1:4 unun.

Ruthroff also derived equations for the input impedances seen at either end of the transformer with the opposite end terminated in Z_L. They are:

$$Z_{in} \text{ (low-impedance end)} = Z_0 \left[\frac{Z_L \cos \beta l + j Z_0 \sin \beta l}{2 Z_0 (1 + \cos \beta l) + j Z_L \sin \beta l} \right] \qquad \text{(Eq 1-6)}$$

and

$$Z_{in} \text{ (high-impedance end)} = Z_0 \left[\frac{2 Z_L (1 + \cos \beta l) + j Z_0 \sin \beta l}{Z_0 \cos \beta l + j Z_L \sin \beta l} \right] \qquad \text{(Eq 1-7)}$$

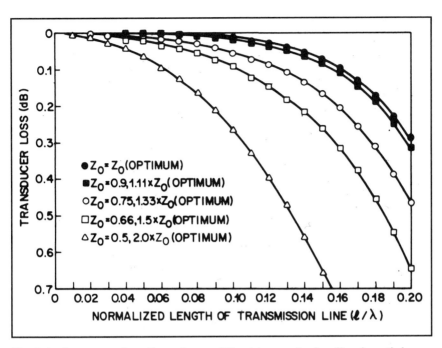

Fig 1-4—Loss as a function of normalized transmission line length in a Ruthroff 1:4 unun for various values of characteristic impedance, Z_0.

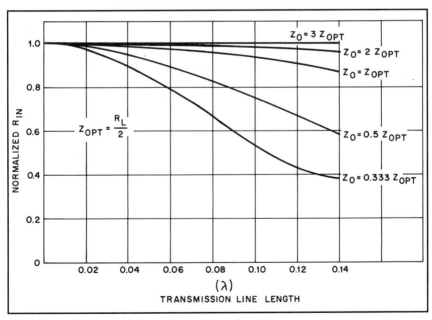

Fig 1-5—The normalized real part of the input impedance of a Ruthroff 1:4 unun as a function of Z_0 and the length of the transmission line.

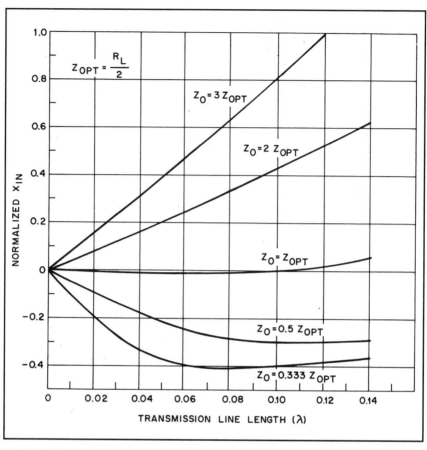

Fig 1-6—The normalized imaginary part of the input impedance of a Ruthroff 1:4 unun as a function of Z_0 and the length of the transmission line.

Pitzalis, et al, plotted Z_{in} (low-impedance end) as a function of various values of Z_0 as compared to the optimum value $Z_0 = 2R_g$ (refs 12, 13). Figs 1-5 and 1-6 are reproductions of their curves for the real and imaginary parts of the input impedance. These curves can be unnormalized by multiplying the ordinate value by $R_L / 4$.

They found that the input impedances were also sensitive to the value of the characteristic impedance. Looking into the low-impedance side of the transformer, the following concepts can be generalized:

    1) For a Z_0 greater than the optimum value:

        a) The real part of Z_{in} increases only slightly with increasing frequency and values of Z_0.

b) The imaginary part of Z_{in} becomes positive and increases with frequency and values of Z_0.

2) For a Z_0 less than the optimum value:

   a) The real part of Z_{in} decreases greatly with increasing frequency and decreasing values of Z_0.

   b) The imaginary part of Z_{in} becomes negative and increases in magnitude with frequency and values of Z_0.

The low-frequency model, which is the same as that of the autotransformer, is handled in the conventional way and will be discussed in Chapter 2.

The high-frequency model of Ruthroff's 1:4 balun, shown in Fig 1-3B, adds a direct voltage V_1 to a delayed voltage $-V_2$ as in the basic building block connected as a phase inverter. It can be shown that the high- and low-frequency responses are the same as his 1:4 unun. Two other comments can be made regarding this approach to a 1:4 balun. They are:

1) Unlike Guanella's balun, this one is unilateral; that is, the high-impedance side is always the balanced side.

2) When the center of the load, R_L, is grounded, the high-frequency response is greatly improved. The balun now performs as a Guanella balun which sums two in-phase voltages.

Chapter 2

Low-Frequency Characterization

Sec 2.1   Introduction

C hapter 1 established a foundation for understanding the theory and design principles of the transmission line transformer. The chapter reviewed the work of Guanella and Ruthroff and presented a comparison of their techniques for obtaining a broad-band transformer with a 1:4 impedance transformation ratio. As was noted, Guanella summed the in-phase voltages of two basic building blocks. Ruthroff summed a direct voltage with a delayed voltage from a basic building block connected as a "boot-strap" for his unun (unbalanced-to-unbalanced transformer). For his balun, he summed a direct voltage with a delayed voltage from a basic building block connected as a phase inverter.

This chapter presents analyses for the low-frequency responses of their transformers. As will be shown, the low-frequency response of the Guanella transformer (which is basically a balun) is highly dependent upon where the ground connections are made. This chapter also presents experimental data on various core materials and geometries and, in particular, treats the topic of the rod v the toroid.

Sec 2.2   Low-frequency Analyses of Ruthroff's 1:4 Transformers

Fig 2-1 shows the low-frequency models for the two Ruthroff 1:4 transformers. Fig 2-1A is the schematic for his 1:4 unun and Fig 2-1B for his 1:4 balun. These models represent the cases when the longitudinal reactances of the coiled transmission lines are insufficient and energy is no longer transmitted by a transmission line mode. Fig 2-1A can be recognized as the schematic of a 1:4 autotransformer. Although the analysis presented here is for the 1:4 unun, it can be shown that the 1:4 balun has the same result.

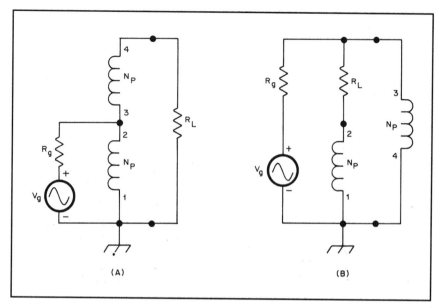

Fig 2-1—Low-frequency models of the Ruthroff 1:4 transformers: (A) unun and (B) balun.

As with the conventional autotransformer, the low-frequency performance of the Ruthroff 1:4 unun can be determined from the reduced model shown in Fig 2-2. Here we have an ideal transformer, shown by the load, labeled R_g, shunted by the core magnetizing inductance, L_M.

If a toroid is used for the core, the magnetizing inductance L_M is

$$L_M = 0.4 \pi N_p^2 \mu_0 \left[\frac{A_e(cm^2)}{l_e(cm)} \right] \times 10^{-8} \text{ henrys} \qquad \text{(Eq 2-1)}$$

where

N_p = the number of primary turns
μ_0 = the permeability of the core
A_e = the effective cross-sectional area of the core
l_e = the average magnetic path length in the core

Eq 2-1 has some important features which should be pointed out here. By making the outside diameter of the toroid as small as is practical, while keeping the same cross-sectional area, a large improvement in bandwidth takes place at both the low- and high- frequency ends of the transformer. With smaller toroids, the length of the transmission

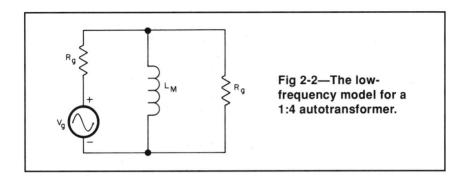

Fig 2-2—The low-frequency model for a 1:4 autotransformer.

line is shorter and the magnetizing inductance is larger because of the shortened average magnetic path length. By using the highest permeability consistent with high efficiency, the low-frequency response is helped further. In fact, by doubling the permeability from 125 to 250, which is a practical value when impedances of less than 200 Ω are involved, the number of turns can be reduced by 30% while maintaining the same low-frequency response. This in turn increases the high-frequency response by 40% since it is inversely proportional to the number of turns. Experiments by the author have shown that toroids with outside diameters between 1.5 and 2 inches can be used in most Amateur Radio applications and still handle the full legal limit of power.

When a rod is used as a transformer core, the calculations for the magnetizing inductance become complicated. This is because of the effect of the high-reluctance air path external to the core. As will be explained shortly, the inductance is independent of the rod's permeability. Experimentally, it is on the order of one-half the value of a toroid with a permeability of 125.

With the following definition for available power,

$$P_{available} = \frac{V_g^2}{4R_g}$$ (Eq 2-2)

the equation for the low-frequency performance of Fig 2-2 can be written as:

$$\frac{P_{available}}{P_{out}} = \frac{R_g^2 + 4X_M^2}{4X_M^2}$$ (Eq 2-3)

where $X_M = 2\pi f L_M$ (Eq 2-4)

It is apparent from Eq 2-3 that the output power approaches the available power when X_M is greater than R_g. Even a factor of five produces a loss of only 1%; the smaller the value of R_g, the smaller the requirement on X_M.

With power transmission line transformers, instead of the well-known 3-dB loss for the upper and lower cutoff frequencies, a more practical figure is 0.45 dB. This represents a loss of about 10% and is equivalent to an SWR of 2:1 when dealing with antennas.

Assuming a loss of 10% at the low-frequency end, the reactance of the primary winding according to Eq 2-3 is:

$$X_M = 3R_g / 2 \qquad \text{(Eq 2-5)}$$

Solving for the number of primary turns using Eqs 2-1, 2-4 and 2-5, we get:

$$N_p \cong \sqrt{\frac{2\,R_g \times 10^7}{f\,\mu_0\,(A_e/l_e)}} \qquad \text{(Eq 2-6)}$$

An approximation is used because some numbers are rounded off and because of small variations in the permeability, μ_0. Experimental results have agreed to within 10% to 20% of the values predicted by Eqs 2-3 and 2-6.

Sec 2.3 Low Frequency Analyses of Guanella's 1:4 Transformers

The low-frequency analysis of the Guanella 1:4 transformer, which is not quite as obvious as the high-frequency analysis, presents some interesting and important results. The low-frequency model is shown in Fig 2-3. It represents the case where energy is no longer transmitted from input to output by a transmission line mode. The low-frequency response is highly dependent upon where the ground connections are made in Fig 2-3 and whether one or two cores (or beaded lines) are employed. With various terminations in Fig 1-2, the Guanella 1:4 transformer can perform as a very wide-band: (A) balun, (B) unun, (C) phase-inverter or (D) hybrid. An understanding of the low-frequency model is important in their designs. The analyses of the more common applications are as follows:

A) *Baluns with floating loads*

Whether terminal 1,5 is grounded, resulting in a step-up balun, or terminal 2 (with the generator now on the high-impedance side) is grounded, resulting in a step-down balun, the low-frequency response is the same in each case. Further, it can be calculated from either side. When looking in at the low-impedance side, one sees the magnetizing inductance L_M, comprised of winding 3-4 in series with winding 6-5. If the two windings are on separate cores, the magnetizing inductance is just the sum of the two separate inductances. But if a single toroidal core is used, and if the windings are in the same direction so as to be series-aiding, thus

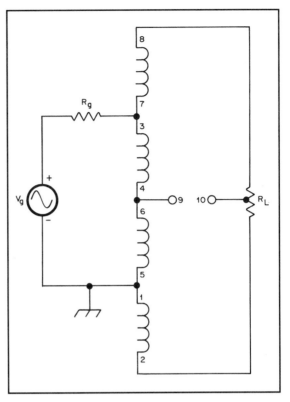

Fig 2-3—Low-frequency model of the Guanella 1:4 transformer.

giving 100% coupling, then the total magnetizing inductance would be greater by a factor of two. It is important to point out here that if the total number of bifilar turns on the Guanella balun (the sum of the two windings) equals the number of turns on the Ruthroff balun, the low-frequency responses are the same, but not the high-frequency responses. The high-frequency response of the Guanella balun is far greater, since in-phase voltages are summed in his technique, and the effective length of the transmission lines is only one-half that of the Ruthroff transformer.

B) *The 1:4 unun (unbalanced-to-unbalanced) transformer*

Since the Guanella transformer adds in-phase voltages, it also offers the best high-frequency response in an unun application. There are two methods for converting the Guanella transformer (which is basically a balun) into unbalanced use: one involves using two separate cores, and the other, adding in series an additional 1:1 balun for isolation. With two grounds, one at terminal 1,5 and the other at terminal 2, it is apparent that a short-circuit exists on winding 1-2 and, hence, also on winding 3-4. If a single core is used, windings 5-6 and 7-8 are also shorted and the low-frequency response is very poor and generally unacceptable. With two separate cores, the inductances of windings 5-6 and 7-8 (which are in series-aiding) determine the low-frequency response. Windings 1-2 and 3-4 only act as a delay line. As was explained before with the basic building block, since no potential difference exists between the input and output terminals of the bottom transmission line in Fig 1-2, the core is no longer needed and can be replaced by a nonmagnetic form for mechanical purposes.

C) *Grounded baluns and hybrids*

Grounding terminals 1,5 and 10 results in the output being balanced to ground. The top of R_L is at $+V_2$ and the bottom at $-V_2$. From Fig 1-2, it can be seen that the input and output voltages on the top transmission line are the same when the line is "flat" ($Z_0 = R_L / 2$). Thus, the top windings have no potential gradients from input to output. Further, the bottom two windings have negative gradients of $-V_2$. This is just the opposite of the unun case, where the bottom transmission line had the zero potential gradient and the top transmission line a positive $+V_2$ gradient. If two cores are used, the top core in Fig 1-2 plays a mechanical role only, since the low-frequency response is now determined by the inductive reactances of windings 1-2 and 3-4. The low- and high-frequency responses are unaffected when a balancing resistor is connected between terminals 9 and 10, as in the symmetrical hybrid case. As in the unun case, if both transmission lines are wound on one toroid, an extra 1:1 balun, in series on the low-impedance side, would be needed for isolation (the ground at terminal 1,5 is, of course, removed).

Sec 2.4 The Rod v the Toroid

What is the feasibility of using rods in place of toroids for broadband transformers? In many cases, because of their simpler form, rods are mechanically preferable. However, the inductance of a rod is more difficult to calculate; much of the magnetic path is in air. Therefore, little quantitative information has previously been available on their use. This section presents experimental results that show rods also

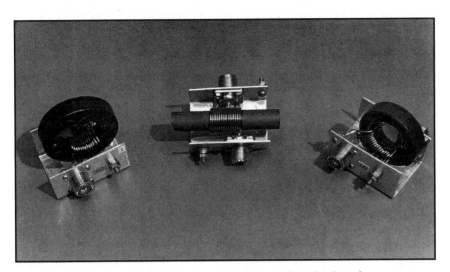

Fig 2-4—Experimental transformers for comparing the low-frequency performance of the rod transformer versus the toroidal transformer. All are tightly wound with 15 inches of no. 14 Formex® wire. The toroidal transformer on the right uses powdered-iron with a permeability of 10. The other two use ferrite cores with permeabilities of 125.

give the same very high efficiencies of the toroid, but at the expense of poorer low-frequency response. Since toroids used in power applications require ferrite permeabilities of only 100 to 300 for very efficient operation (see Chapter 11), the trade-off in using rods is not so severe. The low-frequency response is poorer only by a factor of 2 to 4. In order to establish a comparison between the rod and the toroid, three transformers were constructed with similar windings. Two transformers used toroids of different permeabilities and one used a rod of the same permeability as of one of the toroids. All were tightly wound with 15 inches of no. 14 wire. From early measurements, this was supposed to give a characteristic impedance of 25 Ω, which is considered to be optimum when matching 12.5 Ω to 50 Ω. The three transformers are shown in Fig 2- 4. The toroid on the left had 7 bifilar turns on Q1 ferrite, with a permeability of 125, an OD of 2.4 inches and a thickness of ½ inch. The rod in the center had 7 bifilar turns on Q1 ferrite, a diameter of ⅝ inches and a length of 4 inches. The toroid on the right had 8 bifilar turns on powdered iron (Carbonyl E, Arnold), with a permeability of 10, an OD of 2 inches and a thickness somewhat less than ½ inch, giving rise to one more turn.

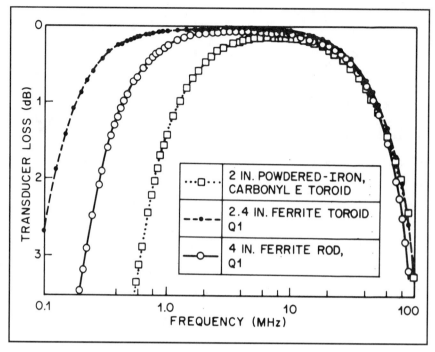

Fig 2-5—Experimental results showing the low-frequency performance of the rod transformer versus the toroidal transformer.

The primary inductances, measured at 1 MHz from the low-impedance side of the transformer with the output open-circuited, were as follows:

Q1 toroid	L = 11.08 μH
Q1 rod	L = 4.67 μH
Powdered-iron toroid	L = 1.67 μH

The experimental results on the loss of these three transformers are shown in Fig 2-5. Several important conclusions can be drawn from these results. They are:

1) The powdered-iron toroid, popular in the Amateur Radio fraternity, is much poorer than the other two. It is not recommended because of its very low permeability.

2) The rod was considerably better than the powdered-iron toroid, and differed from the Q1 toroid only by a factor of about 2. This confirmed the theory of low-frequency response based on the primary inductance measurements. Higher-permeability toroids would give con-

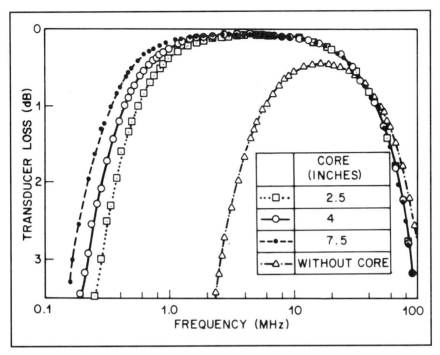

Fig 2-6—Loss measurements vs frequency for four different lengths of rods in a 1:4 transformer.

siderably better low-frequency responses, but cannot be recommended because of excessive core loss. This will be discussed further in Chapter 11.

3) The high-frequency responses were considerably lower than expected. When working with low-impedance matching, in this case 12.5 Ω to 50 Ω, low values of characteristic impedances are necessary. With wire transmission lines, as shown in Fig 2-4, the characteristic impedance can vary considerably, depending upon the spacing between turns. Later measurements of the characteristic impedances showed that they were greater than 25 Ω, giving the lower-than-expected responses.

Sec 2.5   Rod Parameters

As a result of the encouraging comparison of the rod with the toroid, further experiments immediately became apparent. These involved studying the effects of the length and the permeability of the rod on the low-frequency response of the 1:4 transformer.

Three different lengths of Q1 rods (μ = 125), together with no rod at all, were used in the windings of the center transformer in Fig 2-4.

Each rod had a diameter of ⅝ inch. The primary measurements, under the same conditions as in Sec 2.4, produced the following values:

7.5-inch rod	$L_{oc} = 6.07\ \mu H$
4-inch rod	$L_{oc} = 4.67\ \mu H$
2.5-inch rod	$L_{oc} = 3.69\ \mu H$
No rod	$L_{oc} = 0.413\ \mu H$

Fig 2-6 shows the loss measurements v frequency for the four conditions listed above. As expected from the inductance measurements, even a short rod only 2.5 inches long plays an effective role in the low-frequency response. Increasing the length of the rod to 7.5 inches only improves the low-frequency response by about 60%. Therefore, a rod of about 3.5 to 4 inches in length appears to be a practical compromise.

Three rods, each 4 inches in length with a diameter of 0.5 inch, with permeabilities of 125, 260 and 750, were also investigated in a 1:4

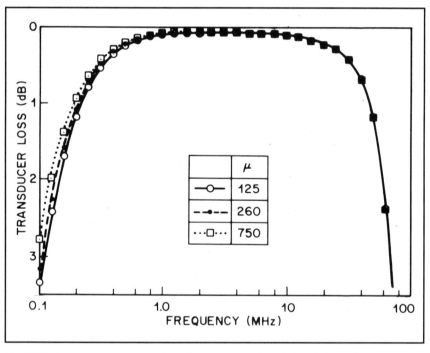

Fig 2-7—Loss measurements v frequency for a 1:4 rod transformer having three different permeabilities. Clearly the low-frequency performance is quite independent of the permeability.

transformer with 10 bifilar turns of no. 14 wire. Fig 2-7 shows the loss measurements v frequency for the three different permeabilities. Notice that there is virtually no difference in the low-frequency response for the three different values of permeability. This experiment clearly showed the strong influence of the high-reluctance air path around the rods.

In summary, the low-frequency performance of a rod transformer using permeabilities of 125 and greater is independent of the permeability, and only mildly dependent on the length of the rod. It compares favorably with a toroidal transformer having a permeability of about 50. Because of the influence of the high-reluctance air path around the core, the efficiency is also quite independent of the permeability. This is not like the case of the toroidal transformer, which has considerably more loss at the permeability levels of 750 and greater, and will be discussed further in Chapter 11. And finally, experiments by the author have shown that rods having diameters of 0.375 inch to 0.5 inch can handle the full legal limit of power in Amateur Radio and are the favorite sizes. It can also be shown mathematically that a 0.375-inch rod with one-third more turns than a 0.5-inch rod has the same low-frequency response.

Chapter 3

High-Frequency Characterization

Sec 3.1 Introduction

A s noted in Chapter 1, the basic building block for all transmission line transformers is the simple, coiled transmission line. Guanella used several lines connected in series-parallel combinations, yielding broadband transformers with impedance ratios of $1:n^2$, where n, the number of bifilar windings, has values of 2, 3, 4 and so on. In order to obtain the maximum high-frequency responses, the characteristic impedance of the transmission lines was required to be R_L / n, where R_L is the high-impedance side of the transformer. Experiments by the author have shown this to be the case and that the high-frequency response then is largely determined by the interwinding capacitance between adjacent bifilar turns. As with the 1:1 balun, if the characteristic impedance is the optimum value and the parasitics are kept to a minimum, Guanella's transformers are quite frequency independent.

Ruthroff used a single basic building block for his 1:4 balun and unun (unbalanced-to-unbalanced) transformers. By connecting the bifilar windings in two different ways, each summing a direct voltage and a delayed voltage via the transmission line, he was able to obtain these two broadband transformers. In most cases these transformers are capable of handling the maximum power level specified for Amateur Radio use in the HF band and beyond. Because of the phase difference in Ruthroff's method of combining voltages, another factor came into the determination of the high-frequency response—the length of the transmission line. In fact, at the frequency at which the electrical length of the transmission line becomes a half-wavelength, the high-frequency response of the transformer is zero. Therefore, short coiled transmission lines, with sufficient inductance to satisfy the low-frequency limit, are required in order to obtain wideband responses from his transformers.

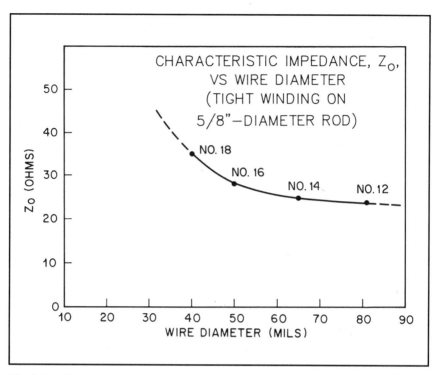

Fig 3-1—Characteristic impedance, Z_0, v wire diameter for a tightly wound transformer.

One of the purposes of this chapter is to compare the theory of Ruthroff with experiments on his 1:4 unun, since considerable information has been published in the literature on this transformer. His technique of bootstrapping (connecting the input to the bottom of the transmission line at the output) is essentially the same as that used by the author to obtain wideband response with ratios less than 1:4 with higher-order windings such as trifilar, quadrifilar and so on. Also included in this chapter is the result of a study on twisted transmission lines, autotransformers v transmission line transformers and, finally, a discussion on how ferrite materials designed for operation up to 10 MHz in conventional transformers can perform very well beyond 100 MHz in transmission line transformers.

Sec 3.2   Experiment v Theory

Ruthroff's analysis of the 1:4 unun, which was reviewed in Chapter 1, predicted that the highest frequency response occurs when the characteristic impedance, Z_0, of the transmission line is the geomet-

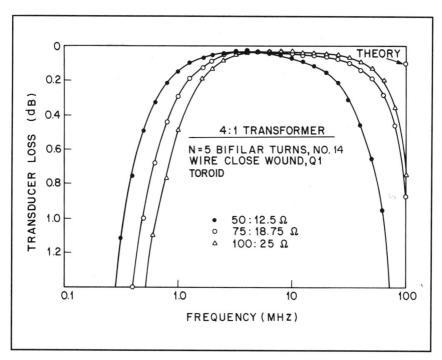

Fig 3-2—Insertion loss v frequency for a tightly wound 4:1 transformer as a function of impedance levels.

ric mean of the input and output resistances. That is, Z_0 should be twice the value of R_g and one-half the value of R_L. Since a need existed for matching the impedance of resonant, short, ground-mounted verticals of about 12 Ω to a 50-Ω coaxial cable, different types of windings using Formex wire were investigated in order to obtain the optimum characteristic impedance of 25 Ω. These early experiments showed that tight windings, as shown in the 1:4 transformers in Fig 2-4, exhibited lower characteristic impedances than twisted pairs. They were also more convenient to use.

Fig 3-1 shows the experimental results of the characteristic impedance of tightly wound, bifilar windings on ⅝-inch diameter rods v wire size. From these data, it appeared that tightly wound no. 14 wire would be best for a 12.5:50-Ω transformer. A transformer was constructed with 5 bifilar turns of no. 14 wire closely wound on a 2.4-inch OD ferrite toroid (permeability of 125). The 5 turns were calculated to give the required low-frequency response.

The results in Fig 3-2 show, however, that the maximum high-frequency response of this transformer occurred at about the 100:25-Ω

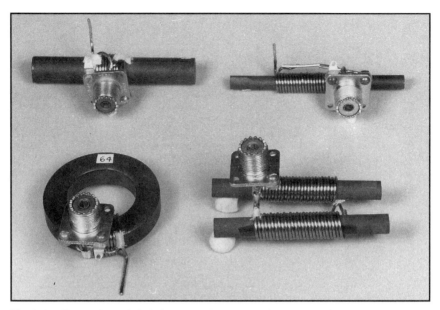

Fig 3-3—Examples of tightly wound 1:4 transformers using no. 14 wire: bottom left, 5 bifilar turns on a no. 64 toroid with a 2.4-inch OD; top left, 5 bifilar turns on a ⅝-inch diameter Q1 rod; top right, 15 bifilar turns on a ⁵⁄₁₆-inch diameter no. 61 rod; and bottom right, a Guanella balun with 12½ bifilar turns on each ⅜-inch diameter no. 61 rod (the other three transformers are Ruthroff ununs).

level instead of the expected level of 12.5:50-Ω. Even then, the experimental value for the high-frequency response was about 70 percent of the theoretical value. In order to understand where the discrepancy in impedance level (nearly a factor of 2) came from, another experiment was undertaken. The characteristic impedance was measured on no. 14 wire, wound very tightly on a toroid before it was connected as an unun. Fig 3-3 shows the toroidal transformer with several other tightly wound rod transformers. Surprisingly, the characteristic impedance was 34 Ω (9 Ω more than expected) and the best high-frequency response occurred at the 17.5:70-Ω level, which is close to the theoretical value. Using other wire diameters again proved the same finding; namely, the characteristic impedances were all considerably higher on toroids than on rods. In fact, some earlier toroidal transformers, with tight windings, were measured and found to have characteristic impedances as high as 41 Ω. Thus there is little doubt that the transformer used in Fig 3-2 had a characteristic impedance in excess of 40 Ω. The reason for the increased value of Z_0 for toroidal transformers is two-fold: first, it is more difficult to get as tight a winding on a toroid as on a rod, and

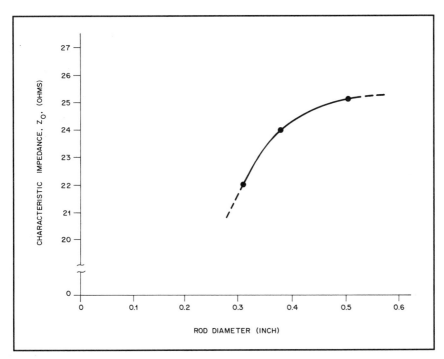

Fig 3-4—Characteristic impedance, Z_0, v diameter for a tightly wound rod transformer using no. 14 wire.

second, the effect on the fringing field is different since the flux is contained within the core.

In the process of winding many practical transformers for the second edition of this book, I found that tightly wound rod transformers did agree quite well with Ruthroff's theory. A characteristic impedance of 25 Ω, using no. 14 wire, did yield the best high-frequency response at about the 12.5:50-Ω level. The question then arises about the results shown in Fig 2-5. Here a ⅝-inch diameter rod had the same high-frequency response as a toroidal transformer! In order to understand this discrepancy, the characteristic impedance was measured on a 5-turn transmission line of no. 14 wire, tightly wound on a ⅝-inch diameter rod. The impedance was found to be 26 Ω (4 Ω higher than expected) and, connected as a 1:4 unun, the best high-frequency response occurred at the 16.5:66-Ω level. Theoretically, the highest frequency response should be at the 13:52-Ω level. Other experiments with rod transformers, when varying the diameter of the rods and the number of turns in order to maintain the same low-frequency response, showed that the characteristic impedance is directly related to the diameter and the

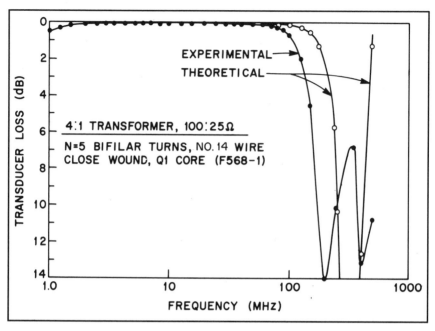

Fig 3-5—Comparison of experimental and theoretical values of insertion loss v frequency for a tightly wound Ruthroff 4:1 transformer.

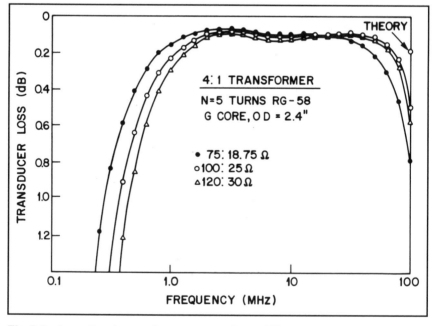

Fig 3-6—Insertion loss v frequency at three different impedance levels for a Ruthroff 4:1 transformer using 50-Ω coaxial cable.

high-frequency response is directly related to the ratio of the length of the coil to its diameter. In other words, the smallest diameter rods gave both the lowest characteristic impedance and the highest frequency response at the 12.5:50-Ω impedance level. Fig 3-4 shows the results of the characteristic impedance measured on rods of varying diameters with closely wound transmission lines of no. 14 wire.

Another interesting comparison between Ruthroff's theory and experiment is shown in Fig 3-5. Here, measurements were made on the same transformer (of Fig 3-2) at the optimum impedance level (100:25-Ω) beyond where the transmission line is a half-wavelength. These data indicate that considerable parasitic coupling exists, introducing a leading component that subtracts from the lagging voltage of a transmission line. This is particularly true at 300 MHz, where an infinite loss should occur instead of the peak, which is only 7 dB down.

Pitzalis and Couse have shown that experiment and theory agree quite well for a 4:1 transformer at the 100:25-Ω level when using a coiled 50-Ω cable (ref 12).[1] I repeated this experiment and the data is presented in Fig 3-6. The results show that the 50-Ω cable is optimum for a 100:25-Ω transformer and that the agreement between theory and experiment is considerably better than with a tightly wound transformer.

A further example of this phenomenon is shown in Fig 3-7. The results presented are for two transformers operating at their optimized impedance levels for the types of windings used. The 40:10-Ω transformer used $\frac{7}{64}$-inch stripline (see Chapter 4) and the 200:50-Ω transformer used no. 18 insulated wire. Fig 3-8 shows the two transformers. Each transformer has the same length of winding and the same number of turns, seven. As expected, the low-frequency performance of the stripline transformer was better by a factor of five. Its series reactance, because of the coiled form of the winding, had to be greater than 10 Ω instead of 50 Ω. But, interestingly, the high-frequency performance of the stripline transformer was only 70% of that of the transformer using no. 18 wire.

Sec 3.3   To Twist or Not to Twist

The discrepancies noted in Sec 3.2 regarding the optimum characteristic impedance and the expected high-frequency performance prompted me to perform further experimental investigations. I com-

---

[1]Each reference in this chapter can be found in Chapter 15.

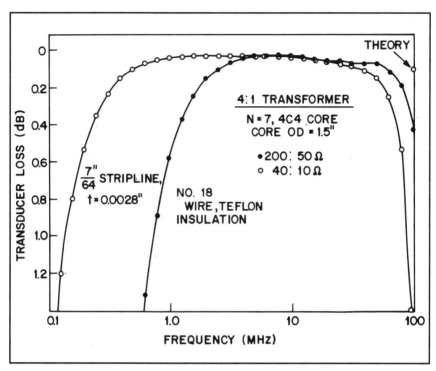

Fig 3-7—Measurements on two transformers with similar lengths of winding, but optimized to operate at 40:10-Ω and 200:50-Ω, respectively.

pared twisted pairs with other types of windings. Twisting has been a popular technique and used by many (refs 9,16). It lowers the characteristic impedance by virtue of the increased capacitance inherent in the close proximity of the wires. The three types of windings investigated were: (1) no. 16 twisted pair (5 turns per inch) with a characteristic impedance of 40 Ω, (2) tightly wound no. 16 wire with a characteristic impedance of 35 Ω, and (3) a pair of no. 16 wires (called twin-lead) held closely together by insulating tape and with a characteristic impedance of 50 Ω. Fig 3-9 shows the three transformers using the various types of windings. By my definition, the center transformer is the tightly wound one. Fig 3-10 shows the experimental results as a function of impedance levels for these 4:1 transformers at A, B, C and D.

On balance, these results favor the simple twin-lead winding. Surprisingly, the twisted pair was no better than the twin-lead winding at the lowest impedance level of 50:12.5- Ω, and the tight winding was the best at the highest level of 120:30-Ω. The poorer low-frequency

Fig 3-8—The two transformers used in the measurements shown in Fig 3-7.

performance of the tightly wound transformer is attributable to the difference in the permeability of its core. This experiment not only points out the effect of the parasitics not considered in the theory, but also the questionable use of twisted pairs for obtaining better low-impedance performance. It should also be noted, however, that Ruthroff used very fine wires (no. 37 and 38 Formex) and extremely small cores (0.175- to 0.25-inch OD). Twisting was necessary to keep the wires closely coupled.

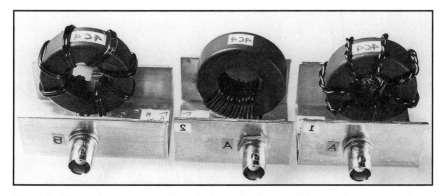

Fig 3-9—Transformers used in the comparison of the performance of twisted pair windings v tightly wound and twin-lead windings.

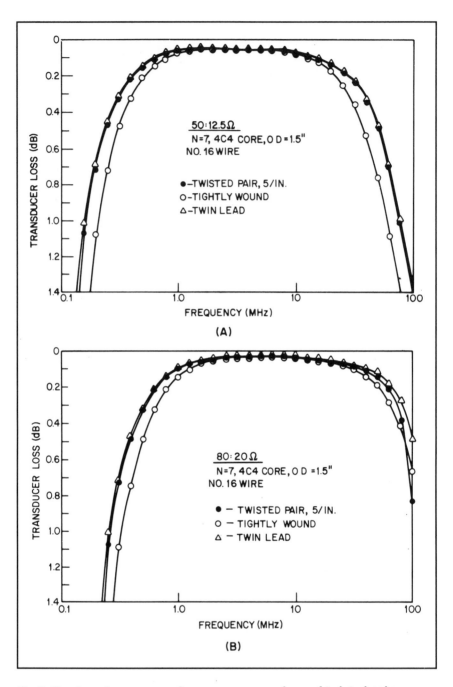

Fig 3-10—A performance vs frequency comparison of twisted pair, tightly wound and twin-lead transformers at four different impedance levels. At A, a comparison of the transformers at 50:12.5-Ω is shown. At B, the comparison is made at 80:20-Ω; in C, the impedance level is 100:25-Ω; and D shows the comparison at 120:30-Ω.

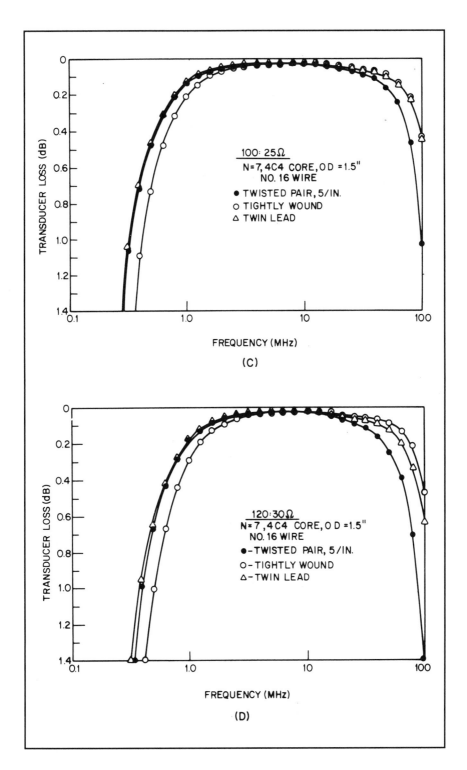

(C)

(D)

Fig 3-11—The three transformers used in comparing the performance of the autotransformer and the transmission line transformer. On the left is an autotransformer. In the center is the transmission line transformer, while on the right is a transmission line transformer without a ferrite core. All transformers had a total of 10 turns.

Sec 3.4   The Autotransformer v the Transmission Line Transformer

The single-winding autotransformer of Fig 2-1A has been around for a long time. In some cases, it has been recommended for use in impedance matching in the HF region (3 to 30 MHz). Little information was available on its comparison with the transmission line transformer, so an experimental investigation was undertaken to determine the differences in performance.

Three 1:4 unbalanced-to-unbalanced transformers were constructed with the same number of turns and types of winding. A photograph of the three transformers is shown in Fig 3-11. The transformer on the left is a 10-turn autotransformer using no. 16 wire wound on a Q1 ferrite toroid ($\mu = 125$) with an outer diameter of 1.25 inch, an inner diameter of 0.75 inch and a core thickness of 0.375 inch. The transformer in the center is a transmission line transformer with 5 bifilar turns of the same wire, also wound on a similar Q1 core. The transformer on the right is a transmission line transformer with the same winding as the center transformer, but without the core. The objective here was to observe, in particular, the effect of the core on the high-frequency response of a transmission line transformer.

Fig 3-12 shows the measurements taken on these transformers when matching 12.5 Ω to 50 Ω. It is immediately apparent that the transmission line transformer is superior to the autotransformer in both

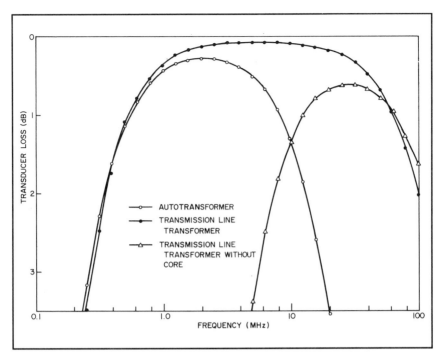

Fig 3-12—Measurements taken on the three transformers in Fig 3-11 when matched at 12.5:50-Ω. Note that the transmission line transformer is superior to the autotransformer both in efficiency and bandwidth.

efficiency and bandwidth. As will be shown in Chapter 11, transmission line transformers frequently exceed 99% in efficiency, while autotransformers rarely exceed 95%. This was an early experiment and the transmission line transformer was not optimized for the 12.5:50-Ω level. If it had been, the differences would be even more dramatic.

Finally, a word about the transmission line transformer without the core. Conventional wisdom states that "the core is needed only for the low-frequency performance where auto-transformer action takes place. At the high-frequency end, the transmission lines are tightly coupled and energy is transmitted by a transmission line mode. Hence a core is not needed. Even a lucite core would work at the high end." Obviously Fig 3-12 and others throughout the book show that the core plays a major role throughout the entire passband of the transformer. Surprisingly, Fig 3-12 also shows that transmission line action is effective even at the low end of the passband (it is even noticeable at 1 MHz).

High-Frequency Characterization   3-13

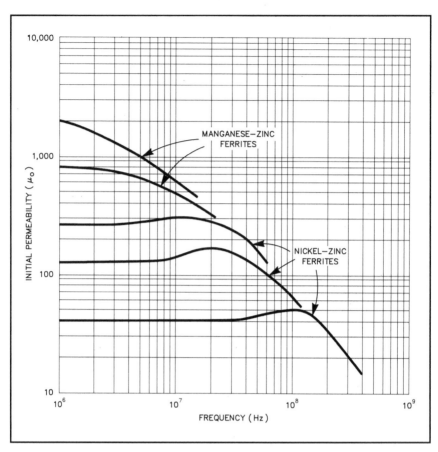

Fig 3-13—Typical frequency responses of nickel-zinc and manganese-zinc ferrites.

Sec 3.5 Ferrites and Frequency Response

The previous section demonstrated that the core still performs a major role at the high-frequency end of a transmission line transformer. The question frequently asked is how can a ferrite material, like Q1, 4C4 or no. 61, designed for operation up to 10 or 20 MHz still produce a flat response beyond 100 MHz? The answer lies in Fig 3-13, which shows that the typical response of most ferrites approaches the same permeability value at the high-frequency end. Since the reactance of a coiled (or beaded) transmission line is proportional to the product of the frequency and permeability (Eqs 2-1 and 2-4) and the slope beyond the knee of the curves in Fig 3-13 shows a near-constant product of permeability and frequency, the following can be said: (1) the maximum reactance of the winding for each ferrite occurs just beyond the knee of the curve, (2) it has a constant value with frequency beyond the knee of the curve, and (3) surprisingly, the maximum values are about the same for all ferrites! Then, one might ask, why not use the highest-permeability ferrite? Fewer turns would be needed at the low-frequency end, which would in turn enhance the high-frequency end (particularly with Ruthroff transformers). The answer to the second question lies in Chapter 11, which deals with materials and power ratings. Briefly, it shows that, in power applications, only nickel-zinc ferrites with permeabilities below 300 produce efficiencies in excess of 98 percent. High-permeability materials, like manganese-zinc ferrite, do not, and are not recommended for power applications.

Chapter 4

Transformer Parameters for Low-Impedance Applications

Sec 4.1 Introduction

More recent experiments by the author (since the publication of the first edition of this book) revealed some interesting results concerning low-impedance applications. Foremost was the design of closely wound rod transformers as compared to toroidal transformers for matching 12.5 Ω unbalanced to 50 Ω unbalanced (matching a short vertical antenna to 50-Ω coaxial cable). Recent measurements showed that, with a tightly wound rod transformer, a characteristic impedance of 25 Ω was easily obtained and the optimum high-frequency response did approach very closely the theoretical prediction by Ruthroff (ref 9).[1]

An early assumption that a tightly wound toroidal transformer had the same characteristic impedance of a rod transformer led to the conclusion that the optimum characteristic impedance for best high-frequency response should be considerably lower than predicted. Actual measurements of the characteristic impedances on toroids (before the transformers were connected as ununs) showed the values to be greater by about 30% over the rods. This is attributed to the difficulty of achieving as tight a winding on a toroid and the different effect of the fringing field on toroidal cores. Transmission loss measurements on toroids again showed that the optimum impedance level for maximum high-frequency response was well predicted by Ruthroff. Some other interesting effects from the length-to-width ratio of the coiled winding on rod transformers were also discussed in the previous chapter.

---

[1]Each reference in this chapter can be found in Chapter 15.

When matching 50-Ω coaxial cable to impedances of 12.5 Ω and lower using toroidal cores, a need arose for transmission lines with characteristic impedances of 25 Ω and lower. These transmission lines could be used both for the Ruthroff transformer with its "boot-strap" connection and for the Guanella transformer with its series-parallel connection. This chapter presents the earlier work of the author on trifilar windings and striplines, as well as the more recent work on transformers using coaxial cable windings. Again, actual measurements of the characteristic impedance, *in situ*, reveal that the maximum high-frequency responses are close to that predicted by Ruthroff and Guanella. The differences are only on the order of 10% for coaxial cable, and as high as 30% for stripline.

Sec 4.2   Stripline Transformers

Lossless transmission lines have a characteristic impedance defined by

Fig 4-1—Four stripline transformers used in the determination of optimum impedance levels for various widths. Each transformer had five bifilar turns on K5 cores of 1.75-inch OD.

$$Z_0 = \sqrt{L / C} \qquad \text{(Eq 4-1)}$$

where

L = the distributed inductance

C = the distributed capacitance.

Z_0 can be lowered by increasing C, by lowering L or by a combination of both. One method to obtain a low value of Z_0 is to use very closely spaced flat conductors called stripline. If the width of the flat conductor is much larger than the spacing between the conductors, the value of the characteristic impedance is

$$Z_0 = 377 \frac{t}{\sqrt{\varepsilon}\ W} \qquad \text{(Eq 4-2)}$$

where

t = the spacing between the conductors

W = the width

ε = the dielectric constant of the insulation.

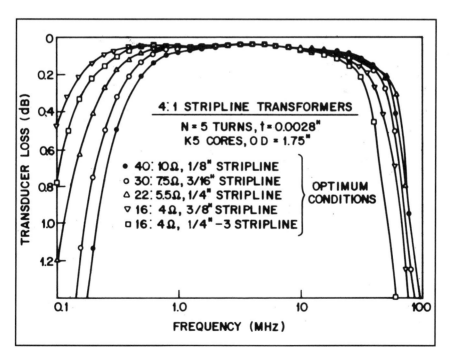

Fig 4-2—Loss v frequency at the optimum impedance levels for the various stripline windings.

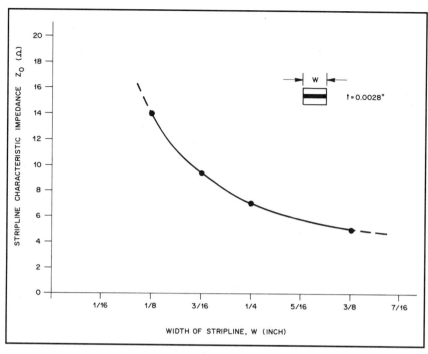

Fig 4-3—Measured values of the characteristic impedance, Z₀, of striplines v width.

To experimentally determine the optimum impedance levels for various widths of stripline, four transformers were constructed and measured at various impedance levels that bracketed the optimum level. The stripline transformers were constructed with widths of ⅛, ³⁄₁₆, ¼ and ⅜ inch. The insulation was Scotch no. 92 tape—a polyimide 2.8 mils thick with excellent electrical, heat-resistant and dielectric properties. The cores were TDK K5 ferrite with an OD of 1¾ inches. Fig 4-1 is a photograph of the four transformers used in the experiment. Fig 4-2 shows the loss-v-frequency curves taken at their optimum levels, that is, impedance levels where the high-frequency response is maximum. Also included in Fig 4-2 is a plot of data for a three-conductor stripline with the third line floating; this will be described further in Sec 4.4. It is interesting to note that the ⅜-inch stripline and the ¼-inch three -conductor stripline have their best high-frequency performance at the low-impedance level of 16:4 Ω. It can be seen from Fig 4-2 that a stripline of about ⁷⁄₆₄ inch in width should be optimum for 50:12.5-Ω operation.

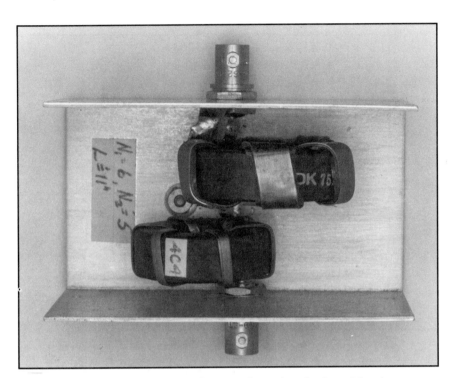

Fig 4-4—Two 4:1 stripline transformers in series designed to give a 16:1 impedance match optimized at the 50:3.125-Ω level.

The four stripline transformers in Fig 4-1 were disconnected as transformers and measured (*in situ*) for their characteristic impedances as simple transmission lines. A plot of the results is shown in Fig 4-3. When relating the measured values in Fig 4-3 to the optimum impedance levels (at the same width) in Fig 4-2, one finds the optimum characteristic impedance to be about 30% lower than predicted by Ruthroff. For example, with ⅛-inch stripline at the 40:10-Ω impedance level (which is optimum), his theory predicted Z_0 to be 20 Ω. Experimentally it was found to be 14 Ω. This percentage difference is the largest obtained from any form of transmission line. Differences from theory with low-impedance coaxial cables are usually less than 10%. With wire transmission lines, the differences are negligible. These differences are probably due to two things: (1) parasitics not included in the theory, and (2) end-effects with short, stripline transmission lines.

Since transmission line transformers exhibit such high efficiencies, it is practical to connect several of them in series to obtain higher transformation ratios. As an example, Fig 4-4 shows two 4:1 transform-

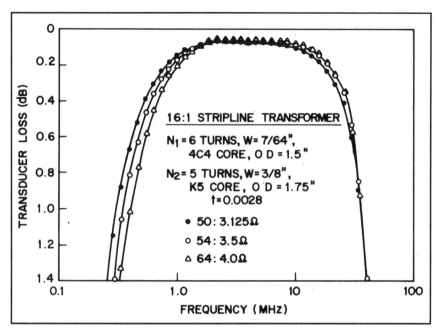

Fig 4-5—Experimental results of the 16:1 transformer of Fig 4-4 at three different impedance levels. Note the best level is at about 54:3.5 Ω.

ers connected in series. They are individually optimized for low-impedance operation. The 4C4 core is wound with six turns of $7/64$-inch stripline designed for matching 50 Ω to 12.5 Ω; the K5 core is wound with five turns of $3/8$-inch stripline designed for matching 12.5 Ω to 3.125 Ω. The insulation is Scotch no. 92 tape. The experimental results in Fig 4-5 show that a considerable portion of the passbands, at the various impedance levels, have less than 0.1 dB of loss—that is, they have better than 98% efficiency. A small improvement in bandwidth could probably be made by using only four turns instead of five on the K5 core with the $3/8$-inch stripline.

Sec 4.3   Low-Impedance Coaxial Cable Transformers

Fig 3-6 shows that a 50-Ω coaxial cable winding presents an optimum match at the 100:25-Ω level and gives a high-frequency response close to that predicted by Ruthroff. To carry this technique to lower impedance levels, a series of various low-impedance coaxial cables were constructed and measured both as straight sections of cables and when wound around toroids. The cables used inner conductors of

Table 4-1

Characteristic Impedance of Coaxial Cables Using Various Combinations of Inner Conductors, Insulators and Outer Braids (Taped)

Insulators and Outer Braids	No. 12 Wire	No. 14 Wire	No. 16 Wire
2 layers of no. 92 tape, RG-122/U braid	12.5 Ω	14 Ω	19.5 Ω
4 layers of no. 92 tape, RG-122/U braid	15 Ω	18.5 Ω	22.5 Ω
6 layers of no. 92 tape, RG-122/U braid	17.5 Ω	21 Ω	26 Ω
2 layers of no. 92 tape and 2 layers of no. 27 (glass tape), RG-122/U braid	21 Ω	23.5 Ω	31 Ω
2 layers of no. 92 tape and 3 layers of no. 27 (glass tape), RG-122/U braid	23 Ω	26 Ω	35 Ω
2 layers of no. 92 tape and 5 layers of no. 27 (glass tape), RG-58/U braid	31 Ω	35 Ω	—

different sizes, insulations of different thicknesses and outer braids that were taped and untaped. The object was to find various characteristic impedances from 10 Ω to 35 Ω. Table 4-1 shows the results when the outer braid is tightly taped (Scotch no. 92 tape is a good choice). These are the values when wound around toroids. When measured as straight sections of coax, the results are about 5% higher. Without taping the outer braid, and when the cables are wound around toroids, the results are about 25% higher than shown in Table 4-1. Obviously, the spacing between the inner conductor and the outer braid is larger under these conditions. For example, a no. 14 wire with two tapes of no. 92 (resulting in four layers of this 2.8-mil thick tape when wound edgewise) becomes 22 Ω without the outer braid being taped, instead of 18.5 Ω. The technique used by the author in constructing these cables is described in detail in Chapter 13.

An example of a 4:1 low-impedance coaxial cable transformer is shown in Fig 4-6. Fig 4-7 shows the schematic. The inner conductor used no. 12 wire and two no. 92 Scotch tapes (wound edgewise, giving

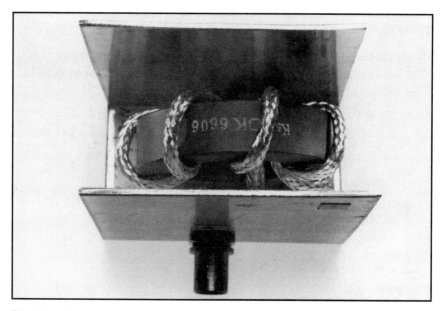

Fig 4-6—A 4:1 transformer using coaxial cable with a characteristic impedance of about 22 Ω.

about four layers). The outer braid, which was untaped, came from RG-58 coax. The transmission loss measurements in Fig 4-8 show the optimum impedance level to be 50:12.5 Ω. This predicts a characteristic impedance of 25 Ω. From Table 4-1, a tightly taped braid with similar inner conductor and insulation yields a value of 15 Ω. For an untaped braid coax, it would increase to about 18.75 Ω (25% higher). Since a larger braid (RG-58 instead of RG-22) was used, and therefore more

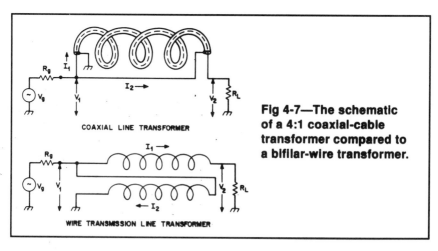

COAXIAL LINE TRANSFORMER

WIRE TRANSMISSION LINE TRANSFORMER

Fig 4-7—The schematic of a 4:1 coaxial-cable transformer compared to a bifilar-wire transformer.

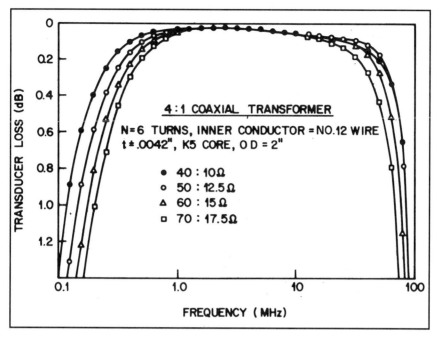

Fig 4-8—Experimental results of a low-impedance coaxial cable transformer as a function of impedance levels. Z_0 = 22 Ω.

spacing occurred between the inner and outer conductors, the characteristic impedance rose to 22 Ω. As in most cases with low-impedance coaxial cables, a value of 10% below the theoretical predictions for the optimum Z_0 was found to be common.

And finally, when designing transformers for low-impedance applications, the currents can be found to be very large at high power levels. Therefore, striplines and coaxial cables have definite advantages over wire lines, since the currents are evenly distributed on the conductors. With wire transformers, the currents are crowded between adjacent turns.

Sec 4.4   The Third Winding

Previous chapters disclosed that bifilar transformers using wire on toroidal cores present a difficulty in obtaining the optimum high-frequency operation at the 50:12.5-Ω level. A characteristic impedance of 25 Ω cannot be achieved with a tight winding. With a rod core it can be easily obtained. In this chapter, striplines and low-impedance coaxial cables have been shown to provide 25 Ω as well as lower values. But

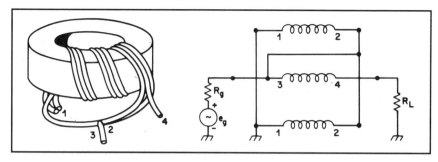

Fig 4-9—A pictorial view and schematic diagram of a 4:1 three- wire transformer with its outer conductors in parallel.

some improvement can be made by using trifilar windings. The result is increased coupling and lower characteristic impedances.

There are two methods of lowering the characteristic impedance with a trifilar winding. The first one uses the two outer wires connected in parallel (Fig 4-9). Fig 4-10 shows the experimental data for a 4:1 transformer with four trifilar turns of no. 14 wire wound on a G core.

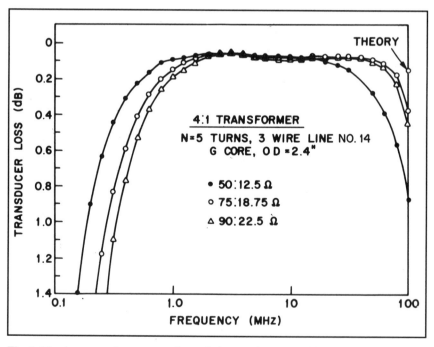

Fig 4-10—Loss vs frequency for a 4:1 transformer with two outer conductors connected in parallel.

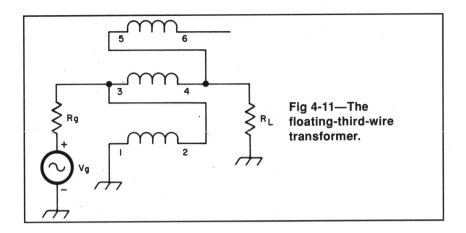

Fig 4-11—The floating-third-wire transformer.

As shown, the transformer is now optimized at the 75:18.75-Ω level. Even the 50:12.5-Ω level performance is much better than that of the close-wound transformer shown in Fig 3-2. Adding more wires in parallel improves the low-impedance level response even further. This

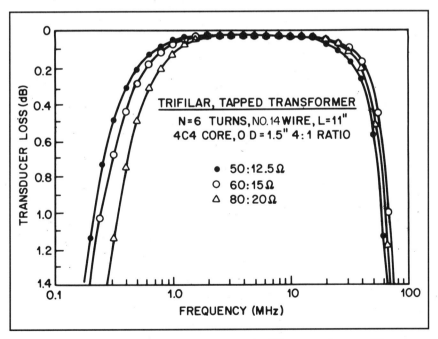

TRIFILAR, TAPPED TRANSFORMER
N=6 TURNS, NO. 14 WIRE, L=11"
4C4 CORE, O D = 1.5" 4:1 RATIO

● 50:12.5 Ω
○ 60:15 Ω
△ 80:20 Ω

Fig 4-12—Experimental results of a tapped trifilar transformer (described in Chapter 8) designed with impedance ratios of 4:1, 5.9:1, 7.36:1 and 9:1. The 4:1 ratio shows the excellent frequency response of a wire transformer at the 50:12.5-Ω level because of the floating third wire.

Fig 4-13—A floating-third-wire transformer. Five trifilar turns of no. 14 wire are wound on a K5 core with an OD of 1.75 inches. An excellent response is obtained at the 50:12.5-Ω level from 0.5 to 70 MHz.

multiwire configuration approaches the performance limit of the coaxial cable.

The second method uses the third wire in a floating connection (Fig 4-11). In this manner, the third wire is left to charge at will and thus lower the characteristic impedance by the increased capacitance effect. This result was found experimentally while observing the performance of 9:1 transformers (input connected to terminal 3 and output to terminal 6; see Chapter 8). The experimental results are shown in Fig 4-12 to demonstrate the dramatic impact of the floating third wire. The results for the 4:1 connection at the 50:12.5-Ω level is extraordinarily good for a wire transformer with a toroidal core. Fig 4-13 shows a floating-third-wire transformer.

As mentioned before, Fig 4-2 also shows the results of another floating-third-wire transformer. The trifilar stripline uses ¼-inch wide strips, with the third strip floating. It compares favorably with the bifilar stripline using ⅜-inch strips. By using a floating strip, the optimum condition for ¼-inch stripline was lowered from the 22:5.5-Ω level to the 16:4-Ω level.

Chapter 5

Transformer Parameters for High-Impedance Applications

Sec 5.1 Introduction

C hapter 4 dealt with transformer parameters for working at low-impedance levels; transformers which are basically used to match the ubiquitous 50-Ω cable to impedances considerably lower than 50 Ω. The need, for the author, first arose when trying to match 50 Ω to the resonant impedance of a short vertical of 12.5 Ω. This required a coiled transmission line with a characteristic impedance of 25 Ω, which was easy to obtain on a rod, but not on a toroid. Various techniques for obtaining low-impedance transmission lines were presented. These included low-impedance coaxial cables, striplines and trifilar combinations.

This chapter deals with parameters for high-impedance applications requiring coiled transmission lines with characteristic impedances of 50 Ω and greater. As will be shown, it is much more difficult to design broadband transformers which can match 50-Ω cable to impedances of 200 Ω and greater. This is especially true of the 1:4 Ruthroff transformers, since impedance ratios are obtained by combining fixed voltages and delayed voltages. Because of the phase differences, his transformers are sensitive to transmission line length. The 1:4 balun of Guanella has the advantage of not only combining voltages of equal delays (thus being somewhat insensitive to the length of the transmission line), but also of combining more than two voltages for even higher impedance ratios. This ability of matching into higher impedances is obtained with the same characteristic impedance as his 1:4 balun.

At higher impedance levels, both Ruthroff and Guanella transformers require many more turns in order to isolate the input from the output and thus discourage the unwanted conventional currents. Further, the spacing between the wires has to be greater in order to achieve the higher characteristic impedances. Eventually, one just runs out of room on a toroid. It becomes physically impossible to wind a high-impedance transmission line transformer with a sufficient number of turns to meet the low-frequency requirement and still have enough space between adjacent turns so as not to lower the characteristic impedance. This situation is made especially difficult when high efficiency is required, since only low-permeability ferrites can be used. Rod core transformers are out of the question in high-impedance applications because of their inability to meet any practical low-frequency requirement.

This chapter, which describes transformer parameters for high-impedance applications, provides the following: (A) curves of actual data on characteristic impedance v spacing and size of wire, (B) a technique using series transformers where a canceling of reactive components occurs, and (C) reviews of the work of others on long transmission lines and variable characteristic impedance lines.

Sec 5.2   High-Impedance Limitations

Ruthroff and Guanella stated that for the maximum high-frequency response of a 1:4 transformer, the characteristic impedance of the coiled transmission line should be one-half the output resistance and twice the generator resistance (Figs 1-2 and 1-3). This results in a flat line since the transmission line is effectively terminated in its characteristic impedance. It is good design practice to try to construct transmission lines that are within 10% of the optimum value. As shown in Fig 1-4 for the Ruthroff transformer, the transmission properties are little affected by Z_0 being off by 10% from the optimum value. For a Guanella transformer, the input impedance, which does not have the sharp cutoff shown in Fig 1-4, would only vary between ±10% (neglecting parasitics).

At the low-frequency end, design considerations focus on the reactance of the coiled transmission line as compared to the impedance level (and hence, the characteristic impedance of the windings). Eqs 2-3 and 2-4 show that if the reactance of the coiled transmission line is only 5 times larger than the generator impedance, R_g, a loss of only 1% occurs. Good design practice requires this reactance to be even greater than 5 times, especially when dealing with the variable impedances of antennas. A conservative design would result in X_M (the reactance of the so-called magnetizing inductance, Eq 2-4) exceeding the characteristic impedance, Z_0, by a factor of about 10. A demonstration of the dependency of the low-frequency response on impedance level is shown in Fig 3-7. Since both transformers have the same number of turns, and hence the same reactance, X_M, the 200:50-Ω level is poorer at the low-frequency end by a factor of 5 over the 40:10-Ω level. Obviously, a 4:1 transformer cannot be used for both high- impedance and low-impedance levels. As in the case of these two transformers, they are either designed for high- or low-impedance operation. It is only when impedance ratios are very low, say around 1.5:1 to 2:1, that transmission line transformers can perform quite well in either direction (for example, matching 50 Ω to a higher or lower impedance). Even then, one direction is usually much better than the other.

High-impedance transmission lines are best obtained with spaced, two-wire lines. The equation for the characteristic impedance for an open-wire line in air is

$$Z_0 = 120 \cosh^{-1} D/d$$

$$= 276 \log 2D/d \text{ for } D \gg d \qquad \text{(Eq 5-1)}$$

where

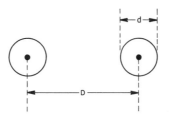

In a dielectric other than air, Eq 5-1 would be reduced by a factor of $1/\sqrt{\varepsilon}$, where ε is the dielectric constant.

As was learned earlier by the author, calculations for the characteristic impedance using a theoretical equation like Eq 5-1 do not produce accurate results in the real world. This is because of the uncertainties in spacing between the conductors, in the effects of the dielectrics, and the proximity effect of neighboring turns. As with the preceding cases of tight wire windings, striplines and coaxial cables on toroids, transformers were wound with various thicknesses of wire and spacings and measured

Fig 5-1—Four of the 22 transformers used in the determination of the characteristic impedance, Z_0, as a function of wire diameter and spacing.

for their characteristic impedances *in situ*. Fig 5-1 shows four of the 22 transformers which were used in obtaining data. The transformer on the lower left uses no. 16 wire spaced about 125 mils by Scotch no. 27 glass tape. Its characteristic impedance is 150 Ω. The transformer in the upper left uses no. 12 wire, with one of the wires having two layers of Scotch no. 92 polyimide tape. Its characteristic impedance is 48 Ω. The transformer in the upper right uses no. 18 wire spaced with a combination of Teflon® tubing and Scotch no. 27 glass tape. Its characteristic impedance is 105 Ω. The transformer in the lower right uses no. 12 wire, with one wire covered with two thicknesses of Scotch no. 27 glass tape. Its characteristic impedance is 63 Ω. The results on the 22 transformers are shown in Fig 5-2. The various spacings were obtained with combinations of thin- and thick-coated wires (1.5 or 3 mils), dielectrics of Scotch no. 92 (2.8 mils), Scotch no. 27 (7 mils) and Teflon tubing with wall thicknesses varying from 12.5 to 25 mils. In some cases, especially with thinner wires, good control of spacing was obtained just with sections of Scotch no. 27 glass tape, as shown in the lower left of Fig 5-1. It should

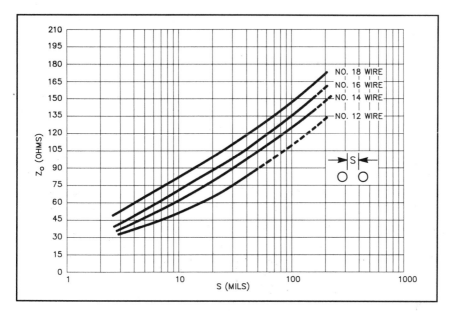

Fig 5-2—Characteristic impedance, Z₀, v wire diameter and spacing, S.

be pointed out that the two smaller spacings of 3 and 6 mils were obtained just with the standard single (thin) and double (thick) coatings on bare copper wire. Also, it can be seen that the plot for the no. 12 wire deviates from a straight line at smaller spacings. This is probably because of the difficulty in obtaining intimate contact between such thick wires. Fig 5-2 also shows that characteristic impedances exceeding 150 Ω are difficult to obtain because of the rather large spacings required between the wires and the effect of neighboring turns. As an example, 300-Ω TV ribbon was wound on a 2.4-inch OD toroid and measurements revealed a characteristic impedance of only 200 Ω.

Another limitation is encountered in the high-impedance use of these transformers, especially where high efficiency is important. As will be shown in Chapter 11, on materials and power ratings, most ferrites exhibit greater loss at higher impedance levels. These losses appear to be dielectric losses, and not magnetic losses as with conventional transformers.

Table 5-1

$P_O/P_{available}$ as a Function of Line Length $(\beta l/2\pi)$ and the Ratio of Load to Source Resistance (R_L/R_g)

$\beta l/2\pi$	1.0	2.0	3.0	4.0	5.0	6.0	9.0	16.0	25.0	∞
0.000	0.640	0.889	0.980	1.000	0.988	0.960	0.852	0.640	0.476	0.000
0.134	0.609	0.847	0.949	0.990	1.000	0.993	0.934	0.768	0.610	0.000
0.167	0.590	0.818	0.923	0.973	0.994	1.000	0.973	0.852	0.714	0.000
0.204	0.559	0.768	0.871	0.928	0.962	0.982	1.000	0.964	0.888	0.000
0.227	0.532	0.723	0.820	0.876	0.914	0.939	0.979	1.000	0.988	0.000
0.236	0.520	0.703	0.795	0.850	0.887	0.912	0.956	0.992	1.000	0.000
0.250	0.500	0.667	0.750	0.800	0.833	0.857	0.900	0.941	0.962	1.000
0.333	0.308	0.333	0.324	0.308	0.290	0.273	0.229	0.165	0.121	0.000
0.500	0.000	0.000	0.000	0.000	0.000	0.000	0.000	0.000	0.000	0.000

Sec 5.3   Long Transmission Lines

Blocker presented an analysis which predicted a good match over a broad frequency band for a wide range of impedance levels (ref 21).[1] His analysis was only applicable for impedance ratios equal to or greater than 4:1 and is as follows:

By using Ruthroff's equations for output power, available power and characteristic impedance for maximum high-frequency response, he arrived at:

$$P_{out} = 4(R_g / R_L)(1 + \cos \beta l)^2$$

$$P_{available} = [2(R_g / R_L)(1 + \cos \beta l) + 1]^2 - \sin^2 \beta l \qquad \text{(Eq 5-2)}$$

A perfect match, $P_{out} = P_{available}$, occurs in Eq 5-2 whenever the electrical length of the line βl and the ratio R_L / R_g are related by:

$$\sec \beta l = R_L / 2R_g - 1 \qquad \text{(Eq 5-3)}$$

This condition can be satisfied for any arbitrary value of R_L / R_g that is equal to or greater than 4. At a given frequency, the higher transformation ratios require longer lines for a perfect match. Table 5-1

---

[1]Each reference in this chapter can be found in Chapter 15.

gives the values of $P_{out} / P_{available}$ as a function of line length for values of R_L / R_g ranging from one to infinity. The table assumes that the characteristic impedance of the line is always adjusted to the relationship $Z_0 = \sqrt{R_g R_L}$. Notice from the table that the perfect matches for ratios greater the 4:1 require considerably longer transmission lines. For example, a 16:1 transformer requires a length of 0.227 λ. This is a very long length of transmission line, even at 30 MHz. It certainly is not realizable on a conventional core. However, transformers in the 100-MHz region can utilize this technique.

Sec 5.4 Variable Characteristic Impedance Lines

The use of tapered lines in wide-band matching at very high frequencies is well known (ref 28). Roy showed analytically that it is possible to achieve matching at the lower frequencies of a 4:1 transformer over a wider band than is possible by using a uniform line (refs 23, 24). He considered the exponential line:

$$Z_c(x) = Z_0 e^{(2\tau x / l)} \qquad \text{(Eq 5-4)}$$

where

Z_0 = the characteristic impedance at x = 0

τ = the taper parameter (near unity in most cases)

l = the length of the transmission line.

Since tapering can best be achieved in microstrip form, this interesting technique is best left to the professional with sophisticated resources. To date, no experimental data have been made available.

Irish approached the wide-band matching problem by using a transmission line that varied as a step-function along the length of the line (ref 25). The analysis he presented was for a line of two sections with differing characteristic impedances. The dimensions of the bifilar winding were changed midway along its length. Analytically he showed an extension by about 30% of the useful bandwidth was possible. Again, no experimental data was given.

Sec 5.5 Series Transformers

At high impedance levels, when R_L is greater than 300 Ω, difficulties arise in trying to obtain a characteristic impedance greater than 150 Ω (and have enough turns to satisfy the low-frequency requirement), which is necessary for maximum high-frequency response. But by using two transformers in series, and by designing one of the transformers so that the characteristic impedance of its transmission line is larger than normally required (that is, $Z_0 > \sqrt{R_L R_g}$), a canceling of the reactive components of the two transformers can occur. This results in a greater bandwidth at the high impedance levels, particularly for Ruthroff 1:4 transformers. An example is shown in Chapter 9 in the case of a 12:1 balanced-to-unbalanced transformer designed to match 600 Ω to 50 Ω. A transformer following this design will match a 50-Ω cable to a rhombic antenna.

This canceling effect can be seen from the plots shown in Figs 1- 5 and 1-6 of Ruthroff's Eq 1-6 for the input impedance as seen at the low-impedance end of the transformer.

Chapter 6

1:4 Unbalanced-to-Unbalanced Transformer Designs

Sec 6.1 Introduction

T he 1:4 unbalanced-to-unbalanced (unun) transformer can be said to have had the most attention in the literature from an analysis standpoint (refs 9-27).[1] Credit is given, in large measure, to Ruthroff, who published his classic paper in 1959 (ref 9). The 1:4 unun finds extensive use in solid-state circuits and in many antenna applications when matching ground-fed antennas such as shortened verticals, vertical beams, slopers and inverted Ls, where impedances of 12 to 13 Ω have to be matched to 50-Ω coaxial cable. Even very short verticals, which are used for mobile operation, can approach impedances of 12 Ω because of losses in the loading coils and the less-than-perfect ground systems.

These 1:4 transformers can be designed basically in two ways: (1) the Ruthroff method, which uses a single, coiled transmission line and a feedback (or "boot-strap") connection in order to sum two voltages, and (2) the Guanella method, which sums two voltages by using two coiled transmission lines (and thus equal delays) in a parallel-series connection. The Guanella transformer, which is basically a balun, requires extra isolation as an unun and was discussed in Sec 2.3 of Chapter 2.

Besides these two different methods, the transformers can also come in many forms. These include large and small rods and toroids; transmission lines of wire, coaxial cable or stripline; beaded lines; and low- and high-impedance versions. In all of these cases, several rules of thumb concerning design can be stated: (1) These transformers are

---

1Each reference in this chapter can be found in Chapter 15.

sensitive to impedance levels; they are designed for either high- or low-impedance applications; (2) The high-impedance design (which can mean impedances of only 300 to 400 Ω) is by far the more difficult to construct just because of size limitations; a 50:200-Ω unun requires twice the number of turns of a 12.5:50-Ω unun for the same low-frequency response; (3) For high-efficiency operation, low-permeability ferrites (100 to 300) are to be used; and (4) Rod transformers, even in the Ruthroff method, can find extensive use in the 1.5- to 30-MHz range when used in low-impedance applications; the rod transformer generally requires twice the number of turns of a toroidal transformer.

The purpose of this chapter is to provide many examples of these 1:4 unun designs. In many cases, the ferrite rods used in these examples came from AM loop-stick antennas and filament chokes from Class-B linear amplifiers. These can be found in many flea markets and Amateur Radio junk boxes. These rods generally have a permeability of 125 and are ideal for this use. The toroids in these designs came from many manufacturers. Generally, all of the ferrite materials (with the proper permeability) from the various suppliers have been found to be acceptable. The choice should be made on the basis of availability and price. The one exception is the 4C4 material from Ferroxcube. It is the hardiest of all the ferrites tested by the author and is recommended in applications where damage can come from very high impedances, such as those experienced in antenna tuners. In this case, the magnetizing inductance can become a part of the resonant circuit, thus creating a damaging high flux density in the core.

Sec 6.2 Schematics and Pictorials

Fig 6-1 shows the schematics and pictorials for the 1:4 unun transformer using the single transmission line in the Ruthroff design. The schematics show the generators at the low-impedance side in a step-up operation. Since these transformers are linear and bilateral, the generators could just as well have been placed on the high side (on the right) in a step-down operation. The main consideration is that the characteristic impedance, Z_0, of the coiled transmission line is one-half the value of the high-impedance side and twice the value of the low-impedance side. A coaxial cable representation is not included, but can be readily understood from pictorials (C) and (D). Many photographs shown throughout the text will also help in visualizing the various windings and connections.

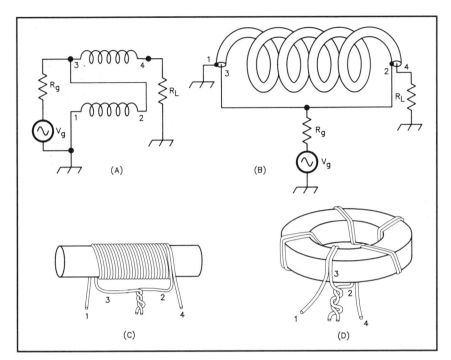

Fig 6-1—The Ruthroff 1:4 unun: (A) wire schematic, (B) coaxial cable schematic, (C) rod pictorial and (D) toroid pictorial.

Fig 6-2 shows the schematics for two versions of the 1:4 unun transformer using the Guanella method of adding two voltages of equal delays. Since the Guanella transformer is basically a balun design, extra isolation has to be considered when operating it as an unbalanced-to-unbalanced transformer. This is explained in Sec 2.3 of Chapter 2. Fig 6-2A shows such an arrangement utilizing a 1:1 balun in series with a 1:4 balun. With sufficient isolation from the 1:1 balun, the input of the 1:1 balun and the output of the 1:4 balun can both be grounded, resulting in very broadband unbalanced-to-unbalanced operation.

The second version basically uses a single core with only the top transmission line in Fig 6-2B wound on it. As was explained in Sec 2.3, when the characteristic impedance of each transmission line has the optimum value, $Z_0 = R_L / 2$, the bottom transmission line has no potential gradient from input to output and therefore requires no magnetic core. The core only gives mechanical support. When the characteristic impedance departs considerably from the optimum value, then winding the bottom transmission line on a magnetic core would improve the

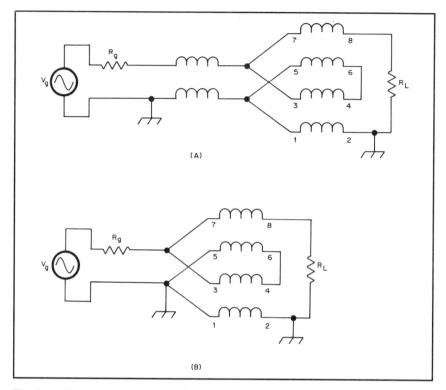

Fig 6-2—Two versions of the Guanella 1:4 unun: (A) a 1:1 balun back-to-back with a 1:4 balun, (B) windings on two separate cores (bottom winding only acting as a delay line).

low-frequency response. The major requirement here is that the reactance of a single winding be much greater than the low-impedance side of the transformer. If the reactances of windings 5-6 and 7-8 in Fig 6-2B have the same values as those of the Ruthroff unun in windings 1-2 and 3-4 in Fig 6-1A, then the two transformers will have the same low-frequency responses. Since Guanella's transformer in Fig 6-2B adds in-phase voltages (windings 1-2 and 3-4 acting only as a delay line), its high-frequency response will be considerably higher than Ruthroff's of Fig 6-1A.

Even though the Guanella ununs are more complicated than that of the single transmission line approach of Ruthroff's, measurements made on these transformers, using the simple test equipment in Chapter 12, show much less phase shift at the higher frequencies, and hence greater high-frequency response. They also lend themselves more readily to hybrid use.

Sec 6.3   12.5:50-Ω Ununs

Fig 6-3 shows five Ruthroff designs (Fig 6-1A) using rods with diameters from ¼ to ⅝ inch. The cable connectors are all on the low-impedance sides of the transformers. The lengths of the rods, which are not critical (see Sec 2.5), are 4 inches. The permeability of each is 125. This is the same as that of the ferrite in the AM loop-stick antenna. The three transformers on the left, which use no. 14 wire, are capable of handling 1 kW of continuous power. The two on the right, which use no. 16 wire, are capable of handling 200 W of continuous power. A single-coated Formvar® (SF) wire has been used successfully. Others, like Formex® and PE (plain enamel), should find equal success. If a more conservative design is needed, then Pyre ML® or H Imideze® (which are the same) are recommended. The latter two have thick coatings (about 3 mils) of an aromatic polyimide. Tightly wound transformers with these polyimide coatings, such as those in Fig 6-3, would have breakdowns similar to those of coaxial cables. Further, the differences in characteristic impedances of the coiled transmission lines due to the extra 1.5-mil thickness have been found to be negligible.

Fig 6-3—Five Ruthroff 1:4 ununs designed to match 12.5 Ω to 50 Ω in the frequency range of 1.5 MHz to 30 MHz. The two on the right are rated at 200 W of continuous power. The three on the left are rated at 1 kW of continuous power.

All of the five rod transformers in Fig 6-3 were optimized for operation at the 12.5:50-Ω level in the 1.5- to 30-MHz range. They have the following parameters:

A) *Upper left:*
5/8-inch diameter, 11 bifilar turns of no. 14 wire.

B) *Middle left:*
1/2-inch diameter, 12 bifilar turns of no. 14 wire.

C) *Lower left:*
3/8-inch diameter, 14 bifilar turns of no. 14 wire.

D) *Upper right:*
5/16-inch diameter, 16 bifilar turns of no. 16 wire.

E) *Lower right:*
1/4-inch diameter, 20 bifilar turns of no. 16 wire.

Toroidal transformers have the advantage of a closed magnetic path, and hence require fewer turns compared to rod versions in order to attain the desired reactance of the coiled transmission line. This shorter length of winding improves the high-frequency response of Ruthroff 1:4 transformers. But these transformers have the disadvantage of not being able to achieve a characteristic impedance of 25 Ω with closely spaced bifilar turns, as can the rod transformers in Fig 6-3. Therefore, other types of transmission lines, such as low-impedance coaxial cable, stripline or floating-third-winding (see Chapter 4), have to be used in 1:4 transformers at the 12.5:50-Ω level and lower. Since stripline is not available to most radio amateurs, transformers using it will not be treated in this chapter. Even though low-impedance coaxial cables are also not readily available, they can easily be fabricated, as is shown in Chapter 13. Coaxial cable and stripline have the advantage in handling higher power levels since the currents are not crowded between adjacent turns as in wire transformers. Further, two or more layers of Scotch no. 92 tape are usually used as the dielectric, thus yielding breakdowns comparable to RG-8/U cable. Many of these coaxial cable transformers are truly in the 5-kW range.

Fig 6-4 shows four versions of 1:4 (12.5:50-Ω) unbalanced-to-unbalanced coaxial cable transformers. The cable connectors are on the low-impedance sides of the transformers. The parameters of these transformers are as follows:

A) *Lower left:*
8 turns of coax using no. 16 wire for inner conductor, insulated with two layers of Scotch no. 92 tape. The outer braid is from RG-174/U cable. The toroid is 1¼-inch OD K5 material (μ= 290).

B) *Upper left:*
8 turns of coax using no. 12 wire for inner conductor, insulated with two layers of Scotch no. 92 tape. The outer braid is from RG-58/U cable. The toroid is 2-inch OD no. 61 material (μ = 125).

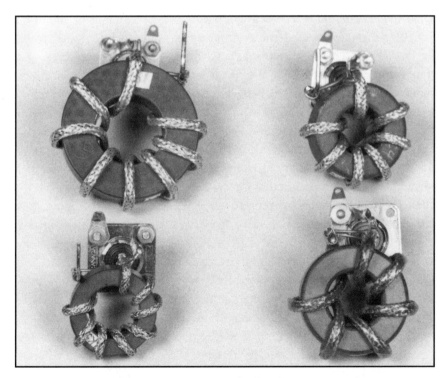

Fig 6-4—Four toroidal versions of the Ruthroff 1:4 unun using low-impedance coaxial cable and designed to match 12.5 Ω to 50 Ω in the frequency range of 1.5 MHz to 30 MHz. The small transformer on the lower left is capable of operating up to 50 MHz.

C) *Upper right:*

7 turns of coax using no. 14 wire for inner conductor, insulated with four layers of Scotch no. 92 tape. The outer braid is from RG-122/U cable. The toroid is 1½-inch OD 4C4 material (μ = 125).

D) *Lower right:*

6 turns of coax using no. 14 wire for inner conductor, insulated with six layers of Scotch no. 92 tape. The outer braid, which is tightly wrapped with Scotch no. 92 tape, is from RG-122/U cable. The toroid is 1½-inch OD 250L material (μ = 250).

The transformer in the lower left of Fig 6-4 has, by far, the widest bandwidth of the four. It ranges from 1.5 MHz to over 50 MHz. This is due to using a smaller toroid and one of the highest permeabilities (μ = 290), while still offering high efficiency. This results in a very short transmission line—only 10½ inches in length. A power test (see Chapter 12) showed that this small transformer was capable of handling 700 W without excessive temperature rise. A conservative rating would be 200 W of continuous power. It is interesting to note that the two transformers on the right in Fig 6-4 use two different types of toroids, inner conductors and outer braids. The characteristic impedances of both coaxial cables are about 22 Ω, even though the one on the lower-right has two more layers of Scotch no. 92 tape. The difference is mainly due to the taping of the outer braid with Scotch no. 92 tape. Wrapping the outer braid with tape reduces the spacing between the inner conductor and outer braid and hence lowers the characteristic impedance by about 25%. Also, since the transformer in the lower-right in Fig 6-4 uses a higher-permeability toroid (250 compared to 125), one fewer turn is required in order to have about the same low-frequency response. This in turn raises the high-frequency response by a ratio of 7/6.

Fig 6-5 shows three versions of a floating-third-winding 1:4 unun using toroids. Without the third winding (see Sec 4.4), the characteristic impedance would be on the order of 45 Ω, and thus have the highest frequency response at the 22.5:90-Ω level. The third winding (see Fig 4-11), which is left floating, reduces the characteristic impedance to about 30 Ω. This enables fairly good 12.5:50-Ω operation. Since the characteristic impedances are in excess of 25 Ω, only small toroids, resulting in the shortest possible lengths of transmission line, are recommended. The parameters of these three transformers are:

Fig 6-5—Three versions of a floating-third-wire 1:4 unun. The transformer on the left is rated at 1 kW of continuous power. The other two are rated at 200 W of continuous power.

A) *Left:*

  7 trifilar turns of no. 14 wire on a 1½-inch OD 4C4 toroid ($\mu = 125$).

B) *Center:*

  9 trifilar turns of no. 16 wire on a 1¼-inch OD Q1 toroid ($\mu = 125$).

C) *Right:*

  8 trifilar turns of no. 16 wire on a 1¼-inch OD toroid ($\mu = 290$).

The larger transformer on the left in Fig 6-5 is capable of handling 1 kW of continuous power. The other two can easily handle 200 W of continuous power. The transformer on the right has the highest permeability, and thus the shortest length of transmission line. This results (with Ruthroff transformers) in the highest frequency response at the 12.5:50-Ω level of the three. The two on the left increase their impedance ratios somewhat (but not at the expense of efficiency) beyond 20 MHz.

The preceding examples in this section used the Ruthroff 1:4 circuit of Fig 6-1A and were capable of covering the 1.5- to 30-MHz range at power levels common to Amateur Radio. These transformers used a single transmission line and, at the 12.5:50-Ω level, allowed for short enough transmission lines so as to satisfy both the low- and high-frequency requirements. The 1:4 Guanella transformers (Fig 6-2) have two transmission lines in a parallel-series connection, resulting in the addition of two in-phase voltages and hence much higher frequency capability. The problem with the Guanella transformer is providing sufficient isolation when operating as an unbalanced-to-unbalanced transformer. This is especially true when both transmission lines are wound on the same core. Although this yields the best low-frequency response when operating as a balun, the grounding of both the input and output terminals makes it impractical as an unun.

As shown on Fig 6-2, this isolation can be obtained in two ways: (1) by connecting a 1:1 balun back-to-back with a 1:4 Guanella balun which has both transmission lines on the same core, and (2) by putting the two transmission lines on separate cores. In the latter case, only one magnetic core is really needed, since the bottom transmission line in Fig 6-2B has no longitudinal potential gradient. Windings 5-6 and 7-8 only determine the low-frequency response. If these windings are the same as windings 1-2 and 3-4 in the Ruthroff 1:4 unun of Fig 6-1A, then the Guanella 1:4 unun will not only have the same low-frequency response, but also a much greater high-frequency response. By using coaxial cable widely spaced on a core (so as to minimize parasitics) or beaded, straight coaxial cable, the 1:4 unun transformers of Fig 6-2B should be capable of operating in the VHF and UHF bands.

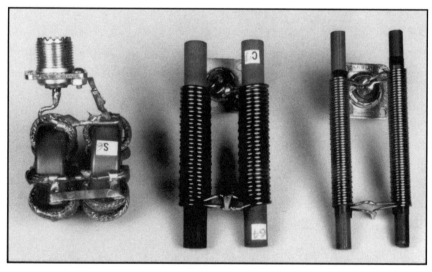

Fig 6-6—Three examples of the very wideband 12.5:50-Ω Guanella unun transformers using the schematic of Fig 6-2(B).

Three examples of these very wideband 12.5:50-Ω Guanella unun transformers, using the schematic of Fig 6-2B, are shown in Fig 6-6. Their parameters are as follows:

A) *Left:*
7 turns of coax on each toroid with no. 14 wire for inner conductor, with six layers of Scotch no. 92 tape. The outer braids, which are tightly wrapped with Scotch no. 92 tape, are from RG-122/U cable. The toroids are 1½-inch OD 250L material ($\mu = 250$).

B) *Center:*
14½ bifilar turns on each rod of no. 14 wire. The rods are ⅜ inch in diameter and are of no. 64 material ($\mu = 250$). Other rods, like Q1 or no. 61 (both $\mu = 125$), would yield similar results (see Fig 2-7).

C) *Right:*
25 bifilar turns of no. 18 wire on ¼-inch diameter rods. The rods are no. 61 material ($\mu = 125$). Actually, the rod on the right has a diameter of ⁷⁄₃₂ inch and two more bifilar turns in order to have a comparable phase delay to the winding on the ¼-inch diameter rod.

The transformers in Fig 6-6 all have their cable connectors on the high-impedance side (50 Ω). Although not shown, in actual operation as a 1:4 unun, one of the low-impedance output leads is grounded. The preference in the coaxial cable version (on the left in Fig 6-6) is to ground the outer braid (the strap connection). The power rating of these three in Fig 6-6 are: on left, 5 kW; in middle, 1 kW; on right, 100 W. These are continuous power ratings.

Sec 6.4   25:100-Ω, 50:200-Ω, and 75:300-Ω Ununs

The 1:4 unun at the 25:100-Ω level presents an interesting case since it probably is the easiest one to design. The optimum characteristic impedance, Z_0, of 50 Ω is readily obtained from RG-58/U or wire with several layers of Scotch no. 92 tape (see Fig 5-2). Further, toroidal transformers with a single transmission line (Ruthroff's design, Fig 6-1A) can still provide ample bandwidths. From the author's experience, there appears to be only a few applications at this impedance level. Most of the useful 1:4 ununs match 50 Ω to either 12.5 Ω or 200 Ω, or 75 Ω to 300 Ω.

Fig 6-7—Four Ruthroff 1:4 ununs (Fig 6-1A) designed to operate at the 25:100-Ω level.

Fig 6-7 shows four Ruthroff ununs designed to operate at the 25:100-Ω impedance level. The cable connectors are all at the low-impedance sides of the transformers. The parameters for these transformers are:

A) *Lower left:*
9 bifilar turns of no. 14 wire on a 1½-inch OD 4C4 toroid (μ = 125). One wire is covered with two layers of Scotch no. 92 tape (Z_0 = 50 Ω).

B) *Upper left:*
10 bifilar turns of no. 14 H Imideze wire on a 1¾-inch OD 250L toroid (μ = 250). (Z_0 = 46 Ω).

C) *Upper right:*
14 bifilar turns of no. 16 wire on a ½-inch diameter rod (μ = 125). One wire is covered with 2 layers of Scotch no. 92 tape (Z_0 = 54 Ω).

D) *Lower right:*

16 bifilar turns of no. 16 wire on a $\frac{5}{16}$-inch diameter rod ($\mu = 125$). One of the wires is conventional hook-up wire with an insulation thickness of 12 mils ($Z_0 = 56 \ \Omega$).

The two toroidal versions in Fig 6-7 easily cover the 1.5- to 30-MHz range with margins. The $\frac{1}{2}$-inch diameter rod version, upper right, can also cover the same frequency range, but with no margin at the low-frequency end. If operation is predominantly in the 1.5- to 7.5-MHz range, then two or three extra bifilar turns should be added. The recommended frequency range of the $\frac{5}{16}$-inch diameter rod transformer (lower right) is from 3 to 30 MHz. At 30 MHz there is considerable phase shift, since the characteristic impedance is higher than optimum. Two layers of Scotch no. 92 tape would present a better thickness of insulation. This smaller rod transformer is capable of handling 200 W of continuous power, and the other three, 1 kW.

A jump in impedance level to 50:200-Ω or 75:300-Ω presents a much more formidable task in design. These higher impedance levels require reactances, due to coiling, to be two and three times greater than the 25:100-Ω series, and four and six times greater than the 12.5:50-Ω series. Therefore, the number of turns has to increase 40% to 70% over the 25:100-Ω series, and 100% to 125% over the 12.5:50-Ω series. Since the high-frequency response is inversely proportional to the length of the transmission line in the Ruthroff transformer (Fig 6-1A), it is very difficult to have high power handling and wideband response with his boot-strap technique. One then has to resort to using a Guanella-based unun, which is not as sensitive to the length of the transmission line since in-phase voltages are summed. Also, since the Guanella transformer is basically a balun, it has to be adapted to unbalanced-to-unbalanced operation, as is shown in Fig 6-2.

Fig 6-8 shows four versions of the Ruthroff transformer capable of working at the 50:200-Ω level with good efficiency. They vary in wideband response and power-handling capability. A description of these four is as follows:

A) *Lower left:*

12 bifilar turns of no. 16 wire on a $1\frac{1}{2}$-inch OD 4C4 toroid ($\mu = 125$). Each wire has 2 layers of Scotch no. 92 tape, resulting in $Z_0 = 75 \ \Omega$. The best level of operation is at 75:150-Ω. Since the trans-

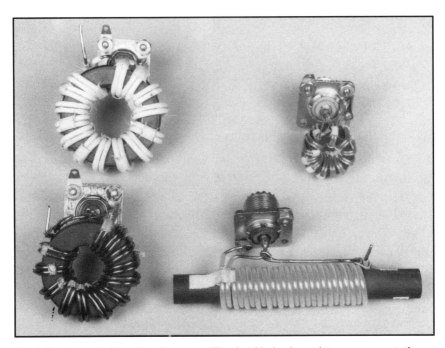

Fig 6-8—Four Ruthroff 1:4 ununs (Fig 6-1A) designed to operate at the 50:200-Ω level.

mission line is quite short (23 inches in length), this transformer can be used quite effectively from the 25:100-Ω level to the 50:200-Ω level. Within this impedance-level range, the transformation ratio is quite constant from 1.5 MHz to 25 MHz. A conservative power rating is 500 W of continuous power. An improved transformer would use 10 bifilar turns on a toroid with a permeability of 250.

B) *Upper left:*
14 bifilar turns of no. 16 wire on a 1¾-inch OD 250L toroid (µ = 250). The sleeving is 15-mil wall plastic. The impedance ratio is fairly constant at 1:4 from 1.5 MHz to 15 MHz. It becomes greater than 1:5 at 30 MHz. A conservative power rating is 500 W of continuous power. As above, a flatter response would be obtained with fewer turns (12 instead of 14).

C) *Upper right:*
11 bifilar turns of no. 22 hook-up wire on a ⅝-inch OD no. 64 toroid (µ = 250). The insulation is 12-mil wall plastic. At the optimum

impedance level of 50:200-Ω, the frequency range is 1.5 MHz (marginally) to over 65 MHz. The transformation ratio only varies from 1:4 to 1:5 over this wide frequency range. This is a result of the short transmission line—only 13 inches long. The power capability is limited only by the melting of the thin wire. This could be 100 W or more.

D) *Lower right:*
13 bifilar turns of no. 16 wire on a ½-inch diameter no. 61 rod (μ = 125). The insulation is 15-mil wall Teflon tubing. At the optimum impedance level of 50:200-Ω, the useful frequency range is 7 MHz to 25 MHz. At 30 MHz, the impedance ratio rises to 1:5. The power rating is 500 W of continuous power.

In order to attain a much higher frequency response at high power levels, the Guanella-based transformers shown in Fig 6-9 have to be used. A description of these transformers is as follows:

A) *Left:*
A 1:1 balun back-to-back with a 1:4 balun (Fig 6-2A). The 1:1 balun (the smaller toroid) has 11 bifilar turns of no. 14 wire on a 1¾-inch OD 250L toroid (μ = 250). One of the wires has two layers of Scotch no. 92 tape (Z_0 = 50 Ω). The 1:4 balun has two windings of 9 bifilar turns, in series aiding (wound in the same direction), of no. 14 wire on a 2.4-inch OD no. 64 toroid (μ = 250). The wires are covered with 18-mil wall Teflon tubing. At the optimum impedance level of 55:220-Ω, the transformation ratio is flat from 1.5 MHz to 50 MHz. A conservative power rating is 1 kW of continuous power.

B) *Center:*
This transformer was designed to match 75 Ω to 300 Ω. One of the toroids only acts as a mechanical support for the transmission line (the bottom bifilar winding in Fig 6-2B). Each toroid has 13 bifilar turns of no. 14 wire on a 3-inch OD 4C4 material (μ = 125). The spacing of the wire of approximately ¼ inch yields a characteristic impedance of 150 Ω. In order to obtain this spacing, one toroid has 18-mil wall Teflon tubing on its wires, with this same tubing as a spacer between the covered wires. The other toroid has its wires spaced with sections of Scotch no. 27 glass tape. At these levels of characteristic impedances, the spacing between wires can vary somewhat without any appreciable

Fig 6-9—Three wideband Guanella 1:4 ununs. The transformer on the left, using the schematic of Fig 6-2A, is designed to match 50 Ω to 200 Ω. The transformer in the center, using the schematic of Fig 6-2B, is designed to match 75 Ω to 300 Ω. The transformer on the right, which also uses the schematic of Fig 6-2B, is designed to match 62.5 Ω to 250 Ω.

effect. At the 75:300-Ω level, the variation in the impedance ratio is only 1:4 to 1:4.3 in going from 1.5 MHz to 30 MHz. A conservative power rating is 1 kW of continuous power.

C) *Right:*

Two toroids are used as above. One toroid has 15 bifilar turns of no. 16 wire, and the other 16 turns of the same wire. The extra phase delay of the 16-turn winding is negligible. The windings, which are covered with 18-mil wall Teflon tubing, have a characteristic impedance of 125 Ω. The cores are 2.4-inch OD no. 64 material (μ = 250). At the optimum impedance level of 62.5:250-Ω, a constant impedance ratio exists from 1.5 MHz to well beyond 30 MHz. If no. 14 wire is used, this flat response would occur at the 50:200-Ω impedance level. A conservative power rating with either wire would be 1 kW of continuous power.

In reviewing the differences, at the 50:200-Ω level, between the two types of Guanella transformers shown in Fig 6-9, the advantage probably goes to the one on the left, which uses two baluns back-to-back (Fig 6-2A). This permits a smaller toroid to be used in the 1:1 balun. If the characteristic impedance of the 1:1 balun is 50 Ω, the same as that of the transmission line feeding this transformer, then the 1:1 balun only adds about two feet to the feeder cable. It is essentially transparent.

Chapter 7

Unbalanced-to-Unbalanced Transformer Designs with Impedance Ratios Less Than 1:4

Sec 7.1 Introduction

L ittle information is available on the characterization and practical design of transmission line transformers with impedance ratios of less than 1:4. Investigations reported in the literature have only treated the bifilar winding and its application to obtain impedance ratios of 1:1, 1:4, 1:9 and 1:16. But many applications can be found for efficient, broadband transformers with impedance ratios of 1:1.5, 1:2 and 1:3. Some examples include the matching of 50-Ω coaxial cable to: (A) vertical antennas, inverted Ls and slopers (all over good ground systems), (B) a junction point of two 50-Ω coaxial cables, (C) 75-Ω coaxial cable and (D) shunt-fed towers performing as vertical antennas.

Early work by the author has shown that the tapped-Ruthroff 1:4 transformer can yield low-impedance designs under certain conditions. Their designs and limitations are included in this chapter. More recent work by the author has shown that, by far, a better technique is to use higher-order windings with the Ruthroff "bootstrap" method. This technique leads to further interesting applications such as broadband, multimatch transformers and compound arrangements leading to a new class of baluns: baluns which are capable of matching 50-Ω cable directly to the low-impedance, balanced inputs of Yagi antennas and to the higher-impedance, balanced inputs of quad antennas.

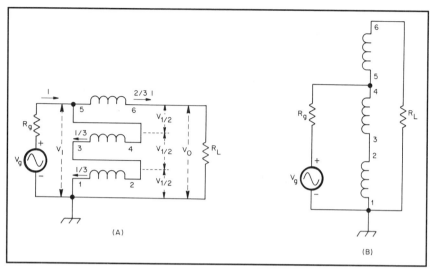

Fig 7-1—Schematics for the 1:2.25 unbalanced-to-unbalanced transformer: (A) high-frequency model and (B) low-frequency model.

The operation of these single-core, higher-order (trifilar, quadrifilar, and so on) winding transformers can be explained, briefly, by the 1:2.25 unun shown in Fig 7-1. The high-frequency schematic of Fig 7-1A assumes that the characteristic impedance of the two transmission lines (the center winding being common) is at the optimized value of $R_L / 3$ and that the reactance of the coiled transmission line is much greater than $R_L / 3$. Fig 7-1A shows the output voltage to be the sum of two direct voltages of $V_1 / 2$ and one delayed voltage of $V_1 / 2$. This leads, in the midband, to $V_o = 3/2 \, V_1$ and therefore an impedance ratio of 1:2.25. The top winding carries 2/3 I and the bottom two windings, 1/3 I. Thus, the flux in the core is essentially canceled out. Since the windings are in series-aiding (100% mutual coupling), as shown in Fig 7-1B, fewer trifilar turns are required in comparison to the bifilar (1:4) model of Fig 2-1A in order to have the same low-frequency response. Thus, with fewer turns, and hence shorter transmission lines, plus the fact that two direct voltages are added to one delayed voltage, these transformers are capable of much higher frequency responses than the regular 1:4 Ruthroff transformer (Fig 1-3A). It can be shown that the high-frequency improvement is on the order of 2.25 times. The top winding can also be easily tapped, giving a broadband impedance ratio near 1:2. In a way, this transformer is a combination of the methods of Ruthroff and Guanella. Two-thirds of the output voltage is tied directly to the input ("raised by its own bootstrap"), and the other one-third is the delayed output from the top transmission line.

This process can be continued further. For example, a quadrifilar winding yields an output voltage of $V_o = 4/3\ V_1$, and hence an impedance ratio of 1:1.78. A quintufilar winding yields an output voltage of $V_o = 5/4\ V_1$, and hence an impedance ratio of 1:1.56. In fact, a 7-winding (septufilar) transformer has been successfully constructed by the author, yielding a very broadband ratio of 1:1.36. It can also be shown that for the same low-frequency response as the regular 1:4 Ruthroff transformer, the high-frequency responses of the quadrifilar and quintufilar transformers are better by factors of about four and five, respectively. In general, the greater the number of windings, the shorter the transmission lines, and the more the in-phase voltages are summed, and hence the greater the high-frequency response.

This chapter presents many low-ratio designs that use bifilar and higher-order windings. Various kinds of conductors, such as stripline and coaxial cable, are also investigated. The presentation on transformers with more than two windings is unique and should find many uses. Also, a novel concept is presented for rearranging windings for optimally matching 50 Ω to higher or lower impedances.

Sec 7.2   1:1.5 Ununs

There are many ways of obtaining an impedance ratio near 1:1.5 in an unbalanced-to-unbalanced transformer using the Ruthroff method of summing a direct voltage with a delayed voltage (or voltages). In the bifilar case, the top winding is tapped at the appropriate point yielding an output voltage, V_o, nearly equal to 1.25 V_1, where V_1 is the input voltage. Higher-order windings, such as trifilar and quadrifilar, can also have their top windings appropriately tapped for the desired 1.25 V_1 output voltage. Even though trifilar and quadrifilar transformers are better than bifilar transformers, the best choice, by far, is the quintufilar transformer. This transformer has an impedance ratio of 1:1.56 without tapping and possesses a wider bandwidth than the other three. This section treats the tapped-bifilar and quintufilar cases.

Sec.7.2.1   Tapped Bifilar Transformers

In the 1:4 Ruthroff, bifilar, unbalanced-to-unbalanced transformer, two potential gradients exist. They are the normal transverse

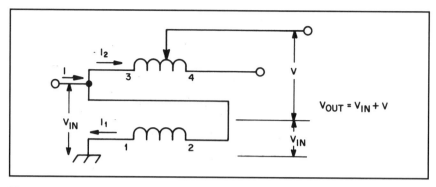

Fig 7-2—Model for the analysis of the tapped bifilar transformer.

potential between the two conductors, and the longitudinal potential from input to output. The transverse potential is related to the flow of energy from input to output. The longitudinal potential merely elevates the output voltage and creates no longitudinal current if the reactance of the coiled winding is sufficient. The performance of the tapped bifilar (and higher-order) transformer is based on this longitudinal gradient.

The model for the analysis is shown in Fig 7-2. With terminal 3 connected to terminal 2, a gradient of V_{in} exists along the length of the bottom winding. If the characteristic impedance of the transmission line is near the optimum value, the same potential gradient will exist along the top winding. No longitudinal currents flow in the windings because of the potential gradients if the reactance of both windings, which are in series-aiding, is much greater than the load, R_L. If the transmission line is wound on a rod, then the tapped voltage, V, which is calculated from terminal 3, is:

$$V = V_{in}l / L \qquad\qquad (Eq\ 7\text{-}1)$$

where

L = the total length of the wire from terminal 3 to terminal 4
l = the length from terminal 3 to the tap.

Thus, the output voltage V_{out} becomes:

$$V_{out} = V_{in}(1 + l / L) \qquad\qquad (Eq\ 7\text{-}2)$$

and the impedance transformation ratio (ρ) becomes:

$$\rho = (V_o / V_{in})^2 = (1 + l / L)^2 \qquad\qquad (Eq\ 7\text{-}3)$$

When $l = L$, ρ has the familiar value of 1:4. Since $V_{in} \times I = V_{out} \times I_2$, it can be shown that the output current becomes:

$$I_2 = I / (1 + l / L) \qquad \text{(Eq 7-4)}$$

With a toroidal core, and with each turn enclosing all of the flux, the gradient can only be tapped at integral turns. Thus, for a toroid, Eq 7-3 becomes:

$$\rho = (1 + n / N)^2 \qquad \text{(Eq 7-5)}$$

where

$\quad$ N = the total number of turns

$\quad$ n = the integral number of turns counted from terminal 3.

Fig 7-3 shows two transformers designed to yield impedance ratios of less than 1:4. The transformer on the right has three turns of five windings, giving a ratio of 1:1.56. This device will be described in Sec 7.2.2. The transformer on the left is a tapped bifilar transformer with seven turns of no. 16 wire on a 1½-inch OD core of 4C4 material ($\mu = 125$). The top winding, shown in Fig 7-2, is tapped at one and three turns from terminal 3. With an output also at terminal 4, the impedance

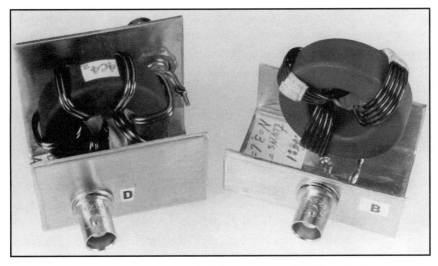

Fig 7-3—The transformer on the left has a tapped bifilar winding yielding ratios of 1:4, 1:2.04 and 1:1.31. The transformer on the right has a 1:1.56 ratio. Both devices use no. 16 wire on a 1.5-inch OD, 4C4 core.

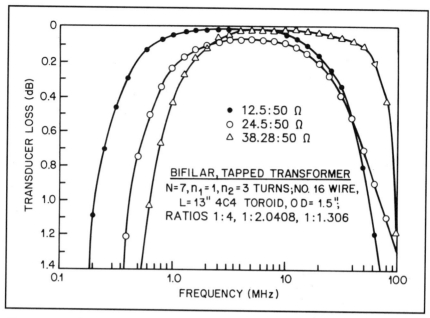

Fig 7-4—Loss v frequency for the tapped bifilar transformer in Fig 7-3.

ratios available are 1:1.31, 1:2.04 and 1:4, respectively. Fig 7-4 shows the loss as a function of frequency for the three different taps when the output is terminated in 50 Ω. Measurements on other tapped-bifilar transformers exhibited similar results.

Fig 7-5 shows two practical tapped-bifilar unun transformers with impedance ratios near 1:1.5. The coax connectors are at the low-impedance sides. Their parameters are as follows:

A) *Left:*

11 bifilar turns of no. 14 wire on a 1½-inch OD K5 toroid ($\mu = 290$). The tap is two turns from terminal 3 in Fig 7-2. The impedance ratio is 1:1.4. At the 38:50-Ω impedance level (which is near optimum), the recommended range is from 1.5 MHz to 15 MHz. Above 15 MHz, the ratio increases and becomes a complex quantity. The power rating is 1 kW of continuous power.

B) *Right:*

14 bifilar turns of no. 16 wire on a ½-inch diameter rod of no. 61 material ($\mu = 125$). The tap is 2¼ turns from terminal 3 in Fig 7-2. The

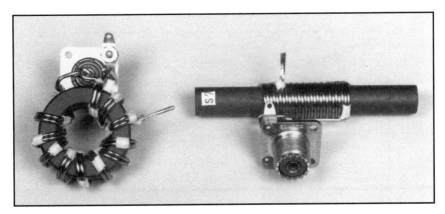

Fig 7-5—Two practical tapped-bifilar unun transformers with impedance ratios near 1:1.5.

impedance ratio is 1:1.35. At the 37:50-Ω impedance level (which is near optimum), the recommended range of operation is from 3.5 MHz to 10 MHz. Above 10 MHz, the ratio decreases and becomes a complex quantity. The power rating is 500 W of continuous power. This transformer is considerably poorer than the toroidal version in Fig 7-5.

Fig 7-6 shows an efficiency curve that tapped bifilar transformers generally exhibit. The lowest efficiency appears at about the 1:2.25 ratio. From this curve, it is apparent that acceptable operation occurs at impedance ratios less than 1:1.5 and greater than 1:3. Also evident is the much poorer performance of the autotransformer, as compared to the transmission line transformer.

Several conclusions can be made concerning tapped bifilar transformers:

1) For the 1:2.04 connection (in Fig 7-4), the loss is considerably greater and the bandwidth considerably less than the other two. The greater loss (also shown in Fig 7-6) suggests that, to a large degree, conventional transformer action is taking place. The transformer should not be operated in this mode. A much better way for obtaining impedance ratios of 1:2 is described in Sec 7.3.

2) The high-frequency response for the 1:1.31 connection (in Fig 7-4) is considerably greater than for that of the 1:4 connection. The effective length of the transmission line is shorter and the characteristic impedance of a transmission line using no. 16 wire favors the

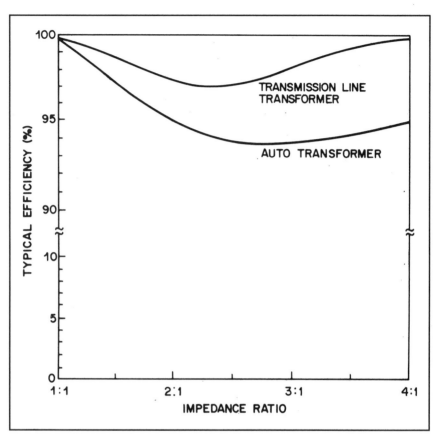

Fig 7-6—A comparison of the efficiency between tapped bifilar transmission line transformers and an autotransformer. Both transformers use no. 16 wire on 4C4 cores of 1.5-inch OD.

38.28:50-Ω level. For many applications, low impedance ratios (around 1:1.3), obtained by tapping, are practical; the transmission line mode prevails and high efficiencies can be obtained. If even greater bandwidths at these low impedance ratios are desired, then going to higher orders of windings is recommended, as is shown in Sec 7.2.2. These higher-order winding transformers can also be tapped, yielding much better bandwidths than the bifilar transformer.

3) Low impedance ratios in the area of 1:1.3 to 1:1.5, with tapped bifilar transformers, require considerably more bifilar turns (for the same low-frequency response) than ratios of 1:3 to 1:4. This is because only a small part of the turns in the top winding in Fig 7-2 play a role in the low-frequency model.

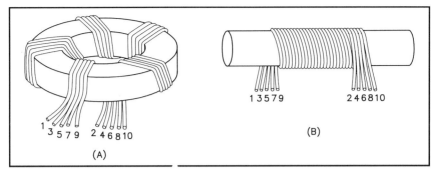

Fig 7-7—Pictorials of quintufilar windings: (A) toroid and (B) rod.

Sec 7.2.2 Quintufilar Transformers

Quintufilar transformers, although being somewhat more difficult to construct, are by far the superior transformer for use around the 1:1.5 impedance ratio. Fig 7-7 shows pictorials for the rod and toroidal versions. As can be seen from Fig 7-7, there is a definite pattern which should help in connecting the various numbers as shown in the schematics which will follow. The photographs of the practical designs will also help.

Fig 7-8 shows schematic diagrams for two versions of the quintufilar, unbalanced-to-unbalanced transformer with an impedance

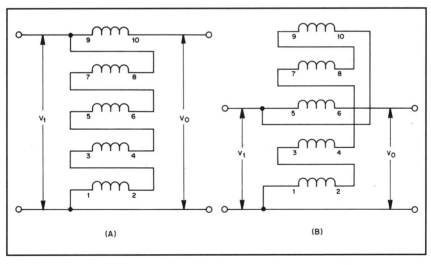

Fig 7-8—Quintufilar transformers with impedance ratios of 1:1.56; A depicts a high-impedance operation, and B displays windings configured for low-impedance operation.

ratio of 1:1.56. Fig 7-8A, using no. 14 or 16 wire wound tightly on a rod (as in Fig 7-7B), gives its highest frequency response when matching 32 Ω to 50 Ω. On a toroid, which cannot be wound as tightly as on a rod, the windings favor matching 50 Ω to 78 Ω. By placing the top winding in Fig 7-8A in the center of the windings as in Fig 7-8B, the characteristic impedance is considerably lowered and optimum performance with a toroidal core can easily occur at the 32:50-Ω level. All of the optimum impedance levels can be easily increased by appropriate thicknesses of insulation on the wires. And finally, it should be noted that these transformers with low impedance transformation ratios, particularly using higher-order windings such as quadrifilar and quintufilar, become quite bilateral in nature. They can be used both as a step-up or a step-down transformer. The main difference is that their high-frequency responses are generally twice as good in the favored direction for which they were designed. In practice, this means a constant transformation ratio up to 45 MHz to 60 MHz in one direction and 25 MHz to 30 MHz in the other.

Fig 7-9 shows the loss as a function of frequency for the five-winding transformer in Fig 7-3. The best high-frequency response

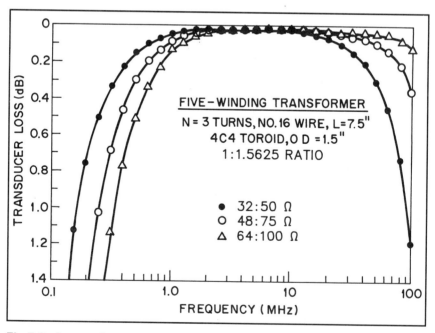

Fig 7-9—Loss v frequency for a five-winding transformer.

occurs at the 64:100-Ω level. Here, a loss of only 0.1 dB extends from 1.2 MHz to 90 MHz! The performance at the 48:75-Ω level also shows very good high-frequency response. For best performance at the 32:50-Ω level, the schematic of Fig 7-8B should be employed.

Fig 7-10 is a photograph of four 1:1.56 rod ununs using the schematic of Fig 7-8A. The two on the left are constructed by first forming a 5-conductor ribbon, held in place every ¾ inch by a Scotch no. 92 clamp. The two on the right are constructed by simply adding one winding at a time (see Chapter 13). The transformer on the bottom left is specifically designed to match 50 Ω to 78 Ω. At this impedance level, the transformation ratio is constant from 1.5 MHz to 40 MHz. At the impedance level of 32:50-Ω, it is constant from 1.5 MHz to 20 MHz. The other three transformers are designed for the 32:50-Ω level. At this level, their impedance ratios are constant from 1.5 MHz to 40 MHz. At the 50:78-Ω level, they are constant from 1.5 MHz to 20 MHz. The parameters for these four transformers, which have the cable connectors at the low-impedance side, are as follows:

A) *Bottom left:*
9 quintufilar turns on a ½-inch diameter rod ($\mu = 125$). The top winding (Fig 7-8A) is no. 14 wire covered with a 10-mil wall plastic

Fig 7-10—Four quintufilar rod ununs. The transformer on the bottom left is designed to match 50 Ω to 78 Ω. The other three are designed to match 50 Ω to 32 Ω.

tubing. The other 4 windings are no. 16 wire. The power rating is 1 kW of continuous power.

B) *Top left:*

9 quintufilar turns on a ½-inch diameter rod (μ = 125). The top winding (Fig 7-8A) is no. 14 wire. The other four windings are no. 16 wire. The power rating is 1 kW of continuous power.

C) *Top right:*

9 quintufilar turns on a ⅜-inch rod (μ = 125). The top winding (Fig 7-8A) is no. 14 wire. The other four windings are no. 16 wire. The power rating is also 1 kW of continuous power.

D) *Bottom right:*

9 quintufilar turns on a ⅜-inch diameter rod (μ = 125). The top winding (Fig 7-8A) is no. 16 wire. The other four windings are no. 18 wire. The power rating is 200 W of continuous power. This transformer can handle 800 W of continuous power with only a small temperature rise.

Toroidal transformers have an advantage over their rod counterparts because of a closed magnetic path and an ability to benefit from higher-permeability ferrites. Rod transformers are insensitive to permeability (see Fig 2-7). As a result, with toroidal transformers fewer turns can be used for the same low-frequency response. This translates to shorter transmission lines and higher frequency responses. Fig 7-11 is a photograph of four 1:1.56 toroidal ununs. The two on the left use 2-inch and 2.4-inch OD cores of no. 61 material (μ = 125) requiring 6 quintufilar turns in order to meet the requirement at 1.5 MHz. The two transformers on the right have smaller cores with higher permeabilities (250 and 290). The two larger toroidal transformers are presented here for readers who have on hand the popular 2- to 2.4-inch OD (μ = 125) ferrite. Their bandwidths are about equal to the four rod transformers in Fig 7-10. From an overall power and bandwidth capability, transformers similar to the one in the upper right in Fig 7-11 are preferred for low impedance ratios at the 1-kW power level. Toroids with outside diameters of 1½ to 1¾ inches and permeabilities of 250 to 300 allow for enough space to wind the appropriate sizes of conductors. Further, this permeability range still offers efficiencies in the 98% to 99% region. The parameters for the four transformers in Fig 7-11 are as follows:

A) *Bottom left:*

6 quintufilar turns on a 2.4-inch OD no. 61 toroid ($\mu = 125$). The schematic of Fig 7-8A is used. The top coil, winding 9-10, has no. 14 wire, and the bottom four, no. 16 each. At the 50:75-Ω level, the impedance ratio is constant from 1.5 MHz to 30 MHz. At the 32:50-Ω level, it is only constant from 1.5 MHz to 15 MHz. The power rating is 1 kW of continuous power.

B) *Top left:*

This transformer is similar to the one above except it uses the schematic of Fig 7-8B, which favors a lower impedance level. It has 6

Fig 7-11—Four quintufilar toroidal transformers. The transformer on the bottom left is designed to match 50 Ω to 78 Ω. The top two transformers are designed to match 50 Ω to 32 Ω. The transformer on the bottom right, although designed for matching 50 Ω to 78 Ω, also works very well in matching 50 Ω to 32 Ω.

quintufilar turns on a 2-inch OD no. 61 toroid (μ = 125). In this case the center coil, winding 5-6, has no. 14 wire, and the other four, no. 16 each. At the 32:50-Ω level, the impedance ratio is constant from 1.5 MHz to 30 MHz. At the 50:75-Ω level, it is constant from 1.5 MHz to 15 MHz. The power rating is 1 kW of continuous power.

C) *Top right*

Only 4 quintufilar turns on a 1½-inch OD no. 64 toroid (μ = 250). The schematic in Fig 7-8B is used. The center coil, winding 5-6, has no. 14 wire, and the other four, no. 16 wire. When matching at the 32:50-Ω level, the impedance ratio is constant from 1.5 MHz to 50 MHz. At the 50:78-Ω level, it is constant from 1.5 MHz to 25 MHz. Since power ratings are more dependent upon conductor sizes and not core sizes, this smaller transformer has the same power rating as the two described above: 1 kW of continuous power. Further, it is an excellent bilateral transformer—stepping down from 50 Ω or stepping up from 50 Ω.

D) *Bottom right:*

5 quintufilar turns on a 1¼-inch OD K5 toroid (μ = 290). The schematic, not shown, consists of interleaving winding 9-10, in Fig 7-8A, between winding 7-8 and winding 5-6. This tends to be a good compromise for a quasi-bilateral transformer. Winding 9-10 has no. 16 wire, and the other four, no. 18. When matching 50 Ω to 78 Ω, the constant impedance ratio is from 1.5 MHz to 100 MHz. At the 32:50-Ω level, it is constant from 1.5 MHz to 50 MHz. A conservative power rating is 200 W of continuous power. This transformer has withstood 700 W without failure.

And finally, Fig 7-12 shows a photograph of three other 1:1.56 unbalanced-to-unbalanced transformers which might prove useful to some readers. The transformer on the left is designed to match (nominally) 100 Ω to 150 Ω. The transformer in the center is designed to match 32 Ω to 50 Ω from 1.5 MHz to 150 MHz. The transformer on the right, a coaxial cable version whose schematic is shown in Fig 7-13, matches 32 Ω to 50 Ω with a 5-kW capability. All connectors are on the low-impedance side of the transformers. Their specific parameters are as follows:

Fig 7-12—Three special quintufilar toroidal ununs. The transformer on the left matches 100 Ω to 150 Ω. The transformer in the center, in matching 50 Ω to 32 Ω, has a 150-MHz capability. The transformer on the right, using low-impedance coaxial cable, has a 5-kW capability.

A) *Left:*

8 quintufilar turns on a 2.4-inch OD no. 67 toroid (μ = 40). The schematic in Fig 7-8A is used. Winding 9-10 uses no. 14 wire covered with a 20-mil wall Teflon tubing. The other four windings are no. 16 wire. At the 100:156-Ω level, the impedance ratio is constant from 1.5 MHz to 30 MHz. The power rating is at least 1 kW of continuous power. For a much higher frequency response, 5 or 6 quintufilar turns on a higher

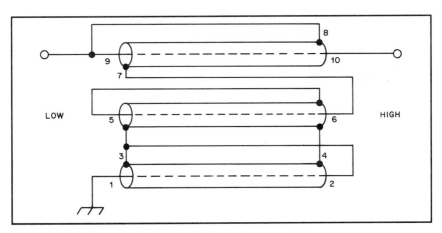

Fig 7-13—Coaxial cable version of quintufilar circuit of Fig 7-8A. The outer braids of the bottom two coaxial cables are connected together.

permeability (250 to 290) and smaller outside diameter (1¾- to 2-inch) toroid is recommended.

B) *Center:*

5 quintufilar turns on a ¾-inch OD no. 64 toroid ($\mu = 250$). Schematic in Fig 7-8B is used. Winding 5-6 uses no. 16 wire. The other four use no. 18 wire. At the 32:50-Ω level, the impedance ratio is constant from 1.5 MHz to 150 MHz. A the 50:78-Ω level, it is constant from 1.5 MHz to 75 MHz. The reasons for this very wideband capability are the use of a relatively high permeability ferrite and the short length of transmission lines—only 5½ inches long. A conservative power rating is 200 W of continuous power. This very small transformer has also withstood 700 W of power.

C) *Right:*

5 quintufilar turns of three coaxial cables connected as a 5-winding transformer (see Fig 7-13 for schematic) on a 2-inch OD K5 toroid ($\mu = 290$). The top coaxial cable has a no. 14 inner conductor with two layers of Scotch no. 92 tape. The outer braid is from RG-122/U cable and is tightly wrapped with Scotch no. 92 tape. The characteristic impedance of this coax is 14 Ω. The other two coaxial cables use no. 16 wire for inner conductors and also have two layers of Scotch no. 92 tape. The outer braids, from RG-122/U cable, are also wrapped with Scotch no. 92 tape. Their characteristic impedances are 19.5 Ω. The highest frequency response occurs at the 45:70-Ω level. At this level, the transformation ratio is constant from 1 MHz to 40 MHz. At the 32:50-Ω level, the ratio is constant from 1 MHz to 30 MHz. Since the current is distributed evenly about the inner conductor of the coaxial cables, a conservative power rating is 5 kW of continuous power. If ML or H Imideze wire were used, the voltage breakdown of this transformer would rival that of RG-8/U cable. Further, by substituting no. 12 wire for the no. 14 wire, and no. 14 wire for the no. 16 wire, a more favorable impedance level would be at 32:50-Ω. Also, the power rating would improve at least two-fold.

Sec 7.3 1:2 Ununs

There are several methods for obtaining transformation ratios around 1:2. Two methods, related to Ruthroff's technique of adding direct voltages to voltages that have traversed coiled transmission lines, are presented in this section. One uses a tapped-trifilar winding, yielding ratios of 1:2.25 and 1:2; the other, a quadrifilar winding yielding a ratio of 1:1.78. Both of these configurations result in very broadband performances since they can employ rather short transmission lines and still satisfy the low-frequency requirements.

Fig 7-1A displayed the circuit diagram for a trifilar design which yields very broadband operation at a 1:2.25 transformation ratio. By connecting the top of the load resistor, R_L, to a tap on the top winding 5-6, an equally broadband (and efficient) ratio can be obtained near 1:2. The tapped version is shown in Fig 7-14. This trifilar transformer is greatly superior to the tapped bifilar transformer in obtaining ratios near 1:2.

Fig 7-15 clearly shows the very high efficiency of the trifilar 1:2.25 (untapped) transformer. Losses of less than 0.1 dB extend over wide bands for three different impedance levels. Transformers with no. 14 wire on toroidal cores have their maximum high-frequency response at about the 44.44:100-Ω level. In this case, the loss is less than 0.04 dB from 1.2 MHz to 30 MHz. This transformer (explained in Chapter 8) also has a 1:9 connection. For a look at the trifilar transformer, see Fig 7-16, upper left.

No. 14 wire yielded the best high-frequency performance at about the 44.44:100-Ω level for the untapped trifilar connection shown in

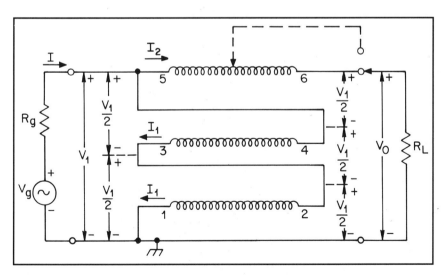

Fig 7-14—Tapped trifilar transformer yielding impedance ratios of 1:1 to 1:2.25.

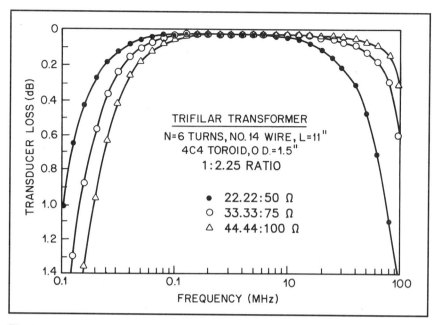

Fig 7-15—Loss v frequency for the trifilar transformer at the 1:2.25 ratio and at three different impedance levels.

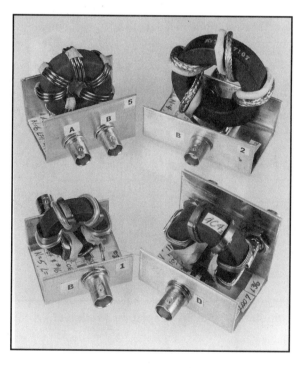

Fig 7-16—Four trifilar-wound transformers using various windings. Looking at the photograph clockwise and starting in the upper left, we have a device made from no. 14 wire, a low-impedance coaxial cable and an insulated third wire, a trifilar stripline, and a bifilar stripline with an insulated third wire.

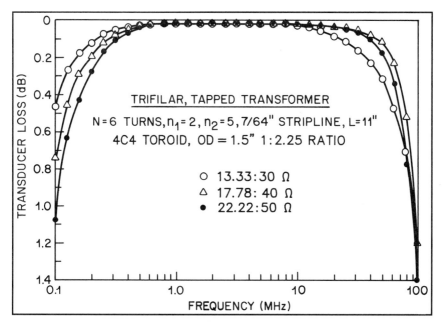

Fig 7-17—Loss v frequency for a tapped-trifilar stripline transformer at the 1:2.25 connection and at three different impedance levels.

Fig 7-14. Therefore, an investigation was undertaken with transmission lines of lower characteristic impedances to obtain better performance at lower impedance levels. A 6-turn, trifilar transformer using $7/64$-inch stripline, with insulation of Scotch no. 92 (2.8 mils thick), was wound on a 4C4, $1\frac{1}{2}$-inch OD toroid. It also had taps at the second and fifth turns from terminal 5 in Fig 7-14. It is shown on the lower right in Fig 7-16. Fig 7-17 shows the performance for the 1:2.25 (untapped connection) ratio. As shown, the response at the 22.22:50-Ω level is very much better than that of the no. 14 windings shown in Fig 7-15. The 0.04-dB loss extends from 0.5 MHz to 30 MHz. It is also evident that this $7/64$-inch trifilar stripline optimizes at the 17.78:40-Ω impedance level. This transformer will also be explained further in Chapter 8.

Two other transmission line configurations that were investigated revealed interesting results. One used a low-impedance coaxial cable for the top two windings in Fig 7-14. The inner conductor of no. 12 wire was insulated with two layers of Scotch no. 92 tape. The bottom winding used an insulated no. 14 wire (seen in the upper right section in Fig 7-16). The performance curves are shown in Fig 7-18. Notice that the

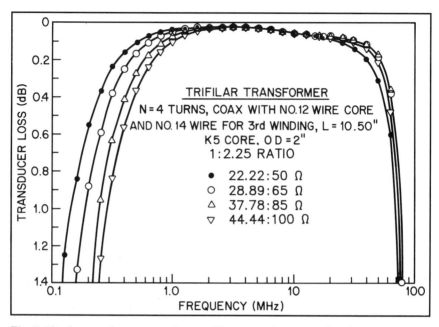

Fig 7-18—Loss v frequency for a trifilar transformer using low-impedance coaxial cable and an insulated third winding at four different impedance levels.

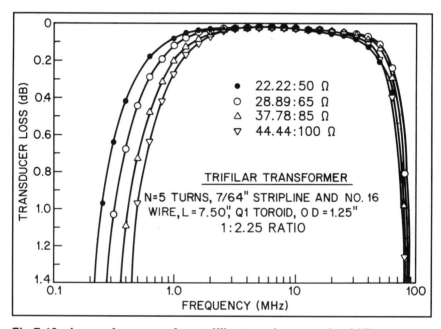

Fig 7-19—Loss v frequency for a trifilar transformer using bifilar stripline and an insulated third wire at four different impedance levels.

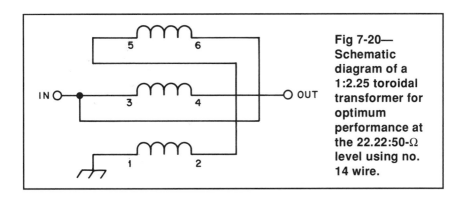

Fig 7-20—Schematic diagram of a 1:2.25 toroidal transformer for optimum performance at the 22.22:50-Ω level using no. 14 wire.

performance at the 22.22:50-Ω level compares quite favorably with the 44.44:100-Ω level of the no. 14 wire transformer. The slight increase in loss above 5 MHz is characteristic of K5 material.

The second transformer used $\frac{7}{64}$-inch stripline with Scotch no. 92 insulation for the top two windings. The bottom winding was an insulated no. 16 wire (transformer shown in lower left portion of Fig 7-16). The performance at the 22.22:50-Ω level, as shown in Fig 7-19, also compares favorably with the 44.44:100-Ω level of the no. 14 wire transformer. The conclusion drawn from the results of these transformers is that the characteristic impedances of the top two windings in Fig 7-14 are most important in determining the high-frequency performance of a trifilar-wound, "boot-strap" transformer.

Probably the most interesting technique for obtaining improved low-impedance operation of a trifilar wire transformer with a toroidal core is the rearrangement of the windings as shown in Fig 7-20. The top winding shown in Fig 7-14 is simply placed in the middle. Here, we have the middle conductor, which carries the larger current, closely coupled to both of the other two windings. This lowers the characteristic impedance and improves the low-impedance performance. Actual data shows that a transformer similar to the one of Fig 7-14, but with the top conductor placed in the center as in Fig 7-20, has a high-frequency response at the 22.22:50-Ω level similar to the response at the 44.44:100-Ω level of Fig 7-14. Thus, when stepping up from 44.44 Ω (also 50 Ω) to 100 Ω, Fig 7-14 is preferred. When stepping down from 50 Ω to 22.22 Ω (25 Ω as well), Fig 7-20 is preferred. Windings of no. 16 wire gave similar results. It should be pointed out that the tightly wound rod transformer presents a completely different case. Because of the tight coupling of the windings, rod transformers favor Fig 7-14 when matching 22.22 Ω to 50 Ω.

An analysis of the tapped trifilar transformer for determining the impedance transformation ratio, ρ, is similar to the bifilar case. For the trifilar rod transformer, the output voltage becomes:

$$V_o = V_1 + V_1 / 2 \times l / L \qquad \text{(Eq 7-6)}$$

where

  l = the length of the transmission line from terminal 5 in Fig 7-14 or from terminal 3 in Fig 7-20
  L = the total length of the transmission line.

The impedance ratio then becomes:

$$\rho = (V_o / V_1)^2 = (1 + l / 2L)^2 \qquad \text{(Eq 7-7)}$$

When $l = L$, the ratio of 1:2.25 is obtained.

For the toroidal transformer, where only integral turns are effective, Eq 7-7 becomes:

$$\rho = (1 + n / 2N)^2 \qquad \text{(Eq 7-8)}$$

where

  n = the number of turns from terminal 5 in Fig 7-14 or from terminal 3 in Fig 7-20
  N = the total number of trifilar turns.

Fig 7-21 shows the outstanding performance of a toroidal transformer tapped at about the 1:2 impedance ratio. The tap is at five turns out of a total of six, from terminal 5 diagrammed in Fig 7-14. As shown, using no. 14 wire optimizes at the 50:100-Ω level. At this level, the loss is less than 0.1 dB from 750 kHz to 75 MHz, and less than 0.04 dB from 1 MHz to 40 MHz. A similar transformer using the schematic diagram of Fig 7-20 gives about the same results, but at the 25:50-Ω level. These two transformers are excellent 1:2 transformers for matching 50 Ω to 100 Ω (Fig 7-14), or 50 Ω to 25 Ω (Fig 7-20). The 50:100-Ω transformer is shown at the right in Fig 7-22.

The trifilar transformer using stripline (shown at the lower right in Fig 7-16) also had taps at two and five turns from terminal 5, Fig 7-14. As mentioned, this transformer was designed for low-impedance

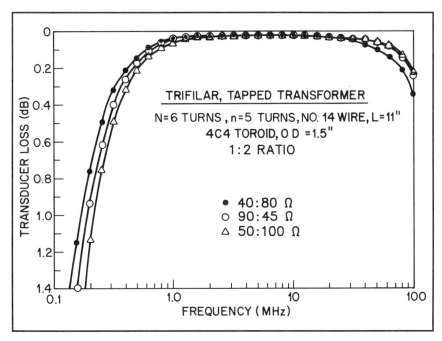

Fig 7-21—Loss v frequency for a tapped trifilar transformer at the 1:2 impedance ratio and at three different impedance levels.

Fig 7-22—The transformer on the right is a tapped trifilar transformer; its performance is shown in Figs 7-15 and 7-21. The transformer on the left is a trifilar transformer designed to give ratios of 1:1.36, 1:2.1, 1:4 and 1:8.2, and is described further in Chapter 8.

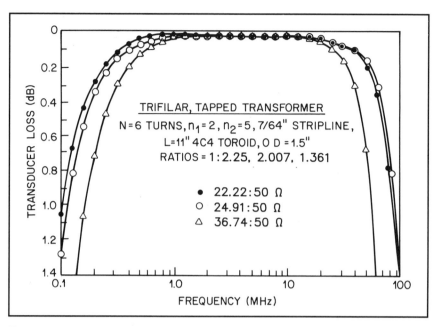

Fig 7-23—Loss v frequency at the 50-Ω output impedance level for a tapped trifilar stripline transformer.

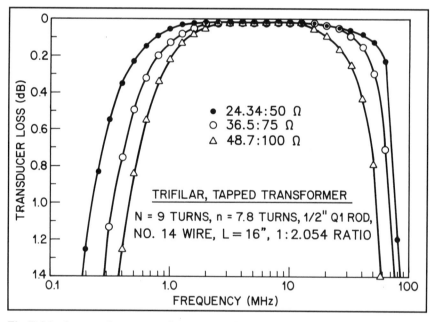

Fig 7-24—Loss v frequency for a tapped trifilar transformer using no. 14 wire on a ½-inch diameter Q1 rod and at three different impedance levels.

operation and had the highest frequency response at the 17.78:40-Ω impedance level when operating as a 1:2.25 transformer (Fig 7-17). With the three output ports, impedance ratios of 1:1.36, 1:2.01 and 1:2.25 are available. Fig 7-23 shows the performance curves for the three different ratios when the high side is terminated in 50 Ω. As shown, true transmission line transformer performance is obtained with the taps at two and five turns. Taps at one, three and four turns gave similar results. From Fig 7-23, it may be seen that the high-frequency performance of the 1:1.36 ratio was not as good as for the other two. This is because the characteristic impedance of the stripline is less than optimum for the 36.74:50-Ω level. In fact, a similar tapped trifilar transformer using no. 14 wire optimizes at the 36.74:50-Ω level.

Fig 7-24 shows the results of a trifilar transformer with nine tightly wound turns of no. 14 wire on a ½-inch diameter Q1 rod (μ = 125) and tapped at 7.8 turns from terminal 5 (Fig 7-14), giving a transformation ratio of 1:2.05. The rod was 3¾ inches long. As shown, optimum high-frequency performance occurs at the 24.34:50-Ω level. Because of the tight winding, as shown on the left in Fig 7-22, rod transformers using the schematic diagrams of Figs 7-14 or 7-20 gave similar results. Although the rod transformer is not quite as good as a toroidal one of similar permeability (Fig 7-23), excellent bandwidth at high efficiency is obtained. At the 24.34:50-Ω level (Fig 7-24), the loss is less than 0.1 dB from 800 kHz to 45 MHz, and less than 0.04 dB from 2 MHz to 25 MHz. This transformer also has transformation ratios of 1:4 and higher and will be described in greater detail in Chapter 8.

Much of the information presented above on the trifilar transformer was essentially reproduced from the first edition of *Transmission Line Transformers*. In order to try to satisfy the requests from many radio amateurs for more practical information, the following part of this section includes descriptions of many other specific designs.

Fig 7-25 shows a photograph of three tapped-trifilar rod transformers with impedance ratios of 1:2 and 1:2.25. They are all designed to handle 1 kW of continuous power. The top transformer is designed to match 50 Ω to 100 Ω or 112.5 Ω. The bottom two transformers are designed to match 50 Ω to 22.22 Ω or 25 Ω. The cable connectors are all on the low-impedance sides of the transformers. Their parameters are as follows:

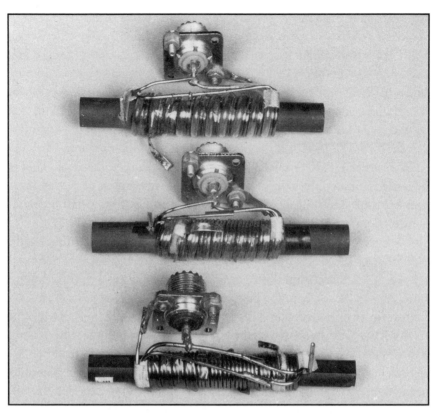

Fig 7-25—Three tapped-trifilar rod transformers with impedance ratios of 1:2 and 1:2.25. The top transformer is designed to match 50 Ω to 100 Ω or 112.5 Ω. The bottom two transformers are designed to match 50 Ω to 22.22 Ω or 25 Ω.

A) *Top:*

10 trifilar turns on a ½-inch diameter, no. 61 rod (μ = 125). The top winding is no. 14 wire, insulated with a 20-mil wall Teflon tubing, and is tapped at 7¾ turns from terminal 5 in Fig 7-14. The other two windings are no. 16 wire. The two impedance ratios of 1:2 (the tapped output) and 1:2.25 are constant from 3.5 MHz to 30 MHz. If operation in the 160-meter band is also desired, then 12 trifilar turns with a tap at 10 turns is recommended.

B) *Middle:*

10 trifilar turns on a ½-inch diameter, no. 61 rod (μ = 125). The top winding is no. 14 wire and is tapped at 8 turns from terminal 5 in

Fig 7-14. The other two windings are no. 16 wire. The two impedance ratios of 1:2 (the tapped output) and 1:2.25 are constant from 1.5 MHz to 45 MHz. The three wires were wound as a ribbon which was held in place by clamps of Scotch no. 92 tape every ¾ inch. This also gave a little spacing between trifilar turns. This optimized the performance at the 22.22:50-Ω and 25:50-Ω levels. This same technique was applied to the other two rod transformers in Fig 7-25.

C) *Bottom:*

13 trifilar turns on a ⅜-inch diameter, no. 61 rod (μ = 125). The top winding is no. 14 wire and is tapped 11 turns from terminal 5 in Fig 7-14. The other two windings are no. 16 wire. The impedance ratios of 1:2 (the tapped output) and 1:2.25 are constant from 1.5 MHz to 45 MHz. As mentioned above, the three wires are wound as a ribbon which is held together by sections of Scotch no. 92 tape. Performance favors, slightly, this transformer over its ½-inch counterpart above.

Fig 7-26 shows a photograph of four tapped-trifilar toroidal transformers with impedance ratios of 1:2 and 1:2.25. The two transformers on the left are designed to match 50 Ω to 22.22 Ω or 25 Ω (the tapped output). The two transformers on the right are designed to match 50 Ω to 112.5 Ω or 100 Ω (the tapped output). The smaller transformer, on the bottom right, is capable of handling 200 W of continuous power. The other three can easily handle 1 kW of continuous power. These transformers have shorter transmission lines than their rod counterparts in Fig 7-25 and, as such, have considerably wider bandwidths. The cable connectors are all on the low-impedance sides of the transformers. Their parameters are as follows:

A) *Bottom left:*

7 trifilar turns of no. 14 wire on a 1½-inch OD, no. 64 toroid (μ = 250). The middle winding in Fig 7-20 is tapped at 6 turns from terminal 3. The impedance ratios of 1:2 (the tapped output) and 1:2.25 are constant from 1 MHz to 45 MHz. This transformer can be used quite well in the reverse direction, that is, matching 44.44 Ω to 100 Ω or 50 Ω to 100 Ω from 1 MHz to 20 MHz.

Fig 7-26—Four tapped-trifilar toroidal transformers with impedance ratios of 1:2 and 1:2.25. The two transformers on the left are designed to match 50 Ω to 22.22 Ω or 25 Ω. The two transformers on the right are designed to match 50 Ω to 100 Ω or 112.5 Ω.

B) *Top left:*

8 trifilar turns on a 1½-inch OD, 4C4 toroid ($\mu = 125$). The middle winding in Fig 7-20 is tapped at 7 turns from terminal 3. The other two windings are no. 16 wire. The impedance ratios of 1:2 (the tapped output) and 1:2.25 are constant from 1 MHz to over 40 MHz. This transformer can also be operated in the reverse direction as above.

C) *Top right:*

8 trifilar turns on a 2-inch OD, K5 toroid ($\mu = 290$). The top winding is no. 14 H Imideze (high breakdown voltage) wire and is tapped at 7 turns from terminal 5 in Fig 7-14. The other two windings are no. 16 wire. For added voltage breakdown margins, all three windings should be H Imideze or ML wire. The impedance ratios of 1:2 (the tapped output) and 1:2.25 are constant from 1 MHz to at least 50 MHz.

D) *Bottom right:*

10 trifilar turns of no. 18 wire on a 1¼-inch OD, K5 toroid (µ = 290). The top winding, in Fig 7-14, is tapped at 9 turns from terminal 5. When matching 50 Ω to 112.5 Ω or to 100 Ω, the impedance ratios are constant from 1 MHz to well over 60 MHz. In the reverse direction, stepping down from 50 Ω, the impedance ratios are constant from 1 MHz to 25 MHz. This transformer, being smaller, has the widest bandwidth of the four in Fig 7-26.

Fig 7-27 is a photograph of two other tapped-trifilar toroidal transformers. The one on the left is made up of two low-impedance coaxial cables with their outer braids connected in parallel and acting as the third conductor. The schematic is shown in Fig 7-28. This transformer, which is optimized to match 50 Ω to 22.22 Ω or 25 Ω, is conservatively rated at 5 kW of continuous power. Also, since it uses four layers of Scotch no. 92 tape as the insulation on the no. 14 inner conductor, its voltage breakdown is similar to that of RG-8/U cable. Further, it demonstrates that the inner conductor of a coaxial cable in a "boot-strap"-connected transformer (Ruthroff) possesses a potential

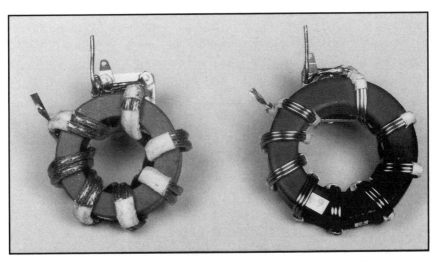

Fig 7-27—Two tapped-trifilar toroidal transformers. The transformer on the left, using coaxial cable and designed to match 50 Ω to 22.22 Ω or 25 Ω, is capable of 5-kW operation. The transformer on the right, designed to match 100 Ω to 44.44 Ω or 50 Ω, uses the popular 2.4-inch OD toroid.

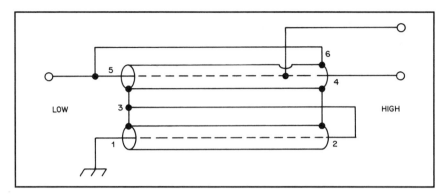

Fig 7-28—The schematic diagram of a tapped-trifilar transformer that uses two sections of coaxial cable, yielding impedance ratios of 1:2 and 1:2.25.

gradient which can be tapped. The transformer on the right is included here since many readers have the popular 2.4- inch-OD toroid (μ = 125) and would find it convenient to use. The parameters for these two transformers are as follows:

A) *Left:*

7 trifilar turns of low-impedance coax (Z_0 = 18.5 Ω) on a 2-inch OD K5 toroid (μ = 290). The no. 14 inner conductors have four layers of Scotch no. 92 tape. The outer braids, which came from RG-122/U cable, are also wrapped with Scotch no. 92 tape. The inner conductor of the top coax in Fig 7-28 is tapped at 6 turns from terminal 5. When matching 50 Ω to 22.22 Ω or to 25 Ω, the impedance ratio is constant from 1 MHz to over 50 MHz.

B) *Right:*

9 trifilar turns of no. 14 wire on a 2.4-inch OD, no. 61 toroid (μ = 125). The top winding in Fig 7-14 is tapped at 8 turns from terminal 5. This transformer matches 100 Ω to 44.44 Ω or 50 Ω with a constant impedance ratio from 1.5 MHz to 40 MHz. The power rating is a conservative 1 kW of continuous power.

Fig 7-29 shows the schematic diagrams for two quadrifilar transformers yielding very broad transformation ratios of 1:1.78. The transformer in Fig 7-29B, because of winding 5-6 being interleaved between windings 7-8 and 3-4, operates better at low impedance levels with

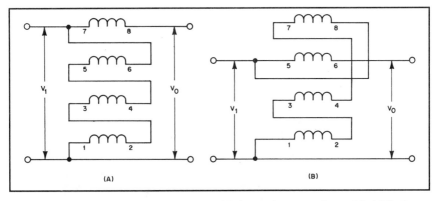

Fig 7-29—Quadrifilar transformers with impedance ratios of 1:1.78; A shows windings for high-impedance operation, and B is configured for low-impedance operation.

toroidal cores. As with trifilar transformers, quadrifilar transformers can also be tapped to give excellent performance at ratios less than 1:1.78. This can best be seen from Fig 7-29A. For a tapped quadrifilar transformer, the output voltage is:

$$V_o = V_1(1 + n / 3N) \tag{Eq 7-9}$$

for a toroidal transformer where

n = the number of turns from terminal 7 in Fig 7-29A and from terminal 5 in Fig 7-29B

N = the number of trifilar turns.

The transformation ratio, ρ, then becomes:

$$\rho = (V_o / V_1)^2 = (1 + n / 3N)^2 \tag{Eq 7-10}$$

Fig 7-30 is a photograph of two quadrifilar transformers with impedance ratios of 1:1.78. They both employ Fig 7-29B, which favors matching 28 Ω to 50 Ω. Their parameters are:

A) *Left:*

6 quadrifilar turns on a 1¾-inch OD, 250L toroid (μ = 250). Winding 5-6 (Fig 7-29B) uses no. 12 H Imideze wire which is covered with two layers of Scotch no. 92 tape. The other three windings are no. 14 H Imideze wire. At the optimum impedance level of 37:65-Ω, the impedance ratio is constant from 1 MHz to 40 MHz. When matching

Fig 7-30—Two quadrifilar transformers with impedance ratios of 1:1.78. They both employ Fig 7-29(B), which favors matching 28 Ω to 50 Ω. The transformer on the left is rated at 2 kW of continuous power. The transformer on the right is rated at 1 kW of continuous power.

50 Ω to 89 Ω (step-up) or 50 Ω to 28 Ω (step-down), the impedance ratio is constant from 1 MHz to 30 MHz. This rather husky transformer, which should be able to handle 2 to 3 kW of continuous power, is quite bilateral.

B) *Right:*

5 quadrifilar turns of no. 14 wire on a 1½-inch OD, 4C4 toroid (μ = 125). This transformer, which uses Fig 7-29(B), is optimized to match 50 Ω to 28 Ω. At this impedance level, the transformation ratio is constant from 1 MHz to over 40 MHz. When matching 50 Ω to 89 Ω, the impedance ratio is constant from 1.5 MHz to 20 MHz. A conservative power rating is 1 kW of continuous power.

Sec 7.4   1:3 Ununs

In many cases the 1:2.25 unun or the 1:4 unun can provide adequate matching when a 1:3 impedance ratio exists. This is particularly true when matching into antennas where the radiation resistance varies with frequency. An example is the resonant quarter-wavelength vertical antenna over a lossless ground system. It has a resistive input impedance of 35 Ω for any reasonable thickness of antenna. This results in a VSWR, at resonance, of 1.4 when feeding directly with 50-Ω coaxial cable. But the lowest VSWR of about 1.25 (and hence best match), occurs a little higher in frequency because of the increased radiation resistance. Resonance and lowest VSWR occur at the same frequency only when the antenna's resonant impedance is 50 Ω (if 50-Ω coax is used) or when a transformer or network matches the impedance of the transmission line to the impedance of the antenna. But on occasion, a 1:3 unun is desirable. An example is the 1:12 balun, shown in Chapter 9, which matches 50 Ω unbalanced to 600 Ω balanced. This balun uses a 1:3 unun in series with a 1:4 balun. Many solid state circuits, which are critical of impedance levels, could also find a 1:3 unun of value.

Two methods have been investigated by the author for obtaining ununs with impedance transformation ratios at, or about, 1:3. One uses the tapped-bifilar schematic shown in Fig 7-2, and the other, a quintufilar schematic similar to those in Fig 7-8. In the quintufilar case, the windings are so interleaved, and the input connections (on the left side) are so made, as to result in two broadband ratios of 1:1.56 and 1:2.78. These transformers are described in Chapter 10, which deals with multimatch transformers. This section describes the tapped-bifilar case.

Fig 7-31 shows a photograph of a tapped-bifilar rod transformer and a tapped-bifilar toroidal transformer. Each has three taps, and

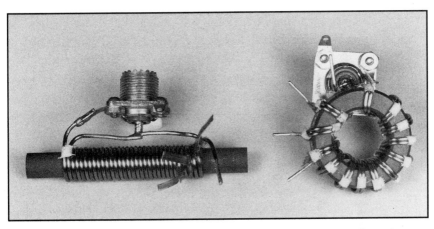

Fig 7-31—Two tapped-bifilar transformers. Each transformer has three taps, yielding four different ratios from 1:3.2 to 1:4. Both transformers are capable of handling 1 kW of continuous power.

therefore four different transformation ratios. They are both capable of handling 1 kW of continuous power. The cable connectors are on the low-impedance sides. The parameters for these two transformers are as follows:

A) *Left:*

16 bifilar turns of no. 14 wire on a ⅜-inch diameter, no. 61 rod (μ = 125). The top winding, in Fig 7-2, is tapped at 9¾, 10⅞ and 12 turns from terminal 3. The author tried for 10, 11 and 12 turns, but wound up a little short on the first two. These differences in turns are insignificant. The impedance ratios are 1:3.2, 1:3.5, 1:3.75 and 1:4, respectively. When matching 50 Ω to 15.6 Ω, 14.3 Ω, 13.3 Ω or 12.5 Ω, all ratios are constant from 1.5 MHz to at least 30 MHz. The maximum high-frequency response varies, from 45 MHz with the 1:4 connection down to 30 MHz for the 1:3.2 connection. Tapping at 8 or 9 turns from terminal 3, for lower impedance ratios, is not recommended.

B) *Right:*

13 bifilar turns of no. 16 wire on a 1½-inch OD, K5 toroid (μ = 290). The top winding, in Fig 7-2, is tapped at 10, 11 and 12 turns from terminal 3. The impedance ratios are 1:3.13, 1:3.41, 1:3.7 and 1:4, respectively. When matching 100 Ω to 29.3 Ω, 27 Ω or to 25 Ω, the impedance ratios are constant from 1 MHz to at least 30 MHz. The 1:4 ratio, which has the highest frequency response, is constant from 1 MHz to 45 MHz. The 1:3.13 ratio, which has the poorest high-frequency response, is constant from 1 MHz to 25 MHz. If the bottom winding in Fig 7-2 is covered with two layers of Scotch no. 92 tape, raising the characteristic impedance of the transmission line to 70 Ω, the transformer would have its optimized performance when matching 150 Ω to 48 Ω, 44 Ω, 40.5 Ω or 37.5 Ω. As above, lower impedance ratios are not recommended.

Chapter 8

Unbalanced-to-Unbalanced Transformer Designs with Impedance Ratios Greater Than 1:4

Sec 8.1 Introduction

Transmission line transformers exhibit exceptionally high efficiencies over considerable bandwidths. By connecting several transformers in series, they can provide practical impedance transformation ratios greater than 1:4. Figs 4-4 and 4-5 in Chapter 4 illustrate operation at a ratio of 1:16, matching 3.125 Ω to 50 Ω with an efficiency of 98% over several octaves. Two Ruthroff 1:4 transformers with low-loss toroids and striplines, with optimized characteristic impedances, were used in this application. Similarly, Chapter 9 includes a balun, stepping up from 50 Ω to 600 Ω, using two Ruthroff 1:4 transformers (one tapped), again optimized for efficient broadband performance. As in the low-impedance example, high efficiency was also obtained over several octaves.

This chapter presents other techniques for obtaining impedance ratios greater than 1:4. They include: converting Guanella's transformers (which are basically baluns) to unbalanced-to-unbalanced operation; adding in series with Guanella baluns higher-order-winding transformers (trifilar, quadrifilar and so on) in order to achieve ratios other than 1:n^2; using fractional-ratio baluns back-to-back with Guanella baluns; applying higher-order windings to the Ruthroff-type transformer; tapping these higher-order Ruthroff-type transformers; and connecting the Ruthroff 1:4 unun in a parallel-series arrangement with a Guanella 1:1 balun.

Guanella's baluns, connected directly as ununs or back-to-back with 1:1 baluns (thus realizing the optimum low-frequency response), offer the greatest bandwidths because they add in-phase voltages. On the other hand, Ruthroff's transformers, which add a direct voltage to a delayed voltage (or to delayed voltages when higher-order windings are used), are much simpler and should find many applications. Further, they can be successfully tapped, yielding a variety of ratios other than 1:4, 1:9, 1:16, and so on. Also, it should become evident from this chapter that high-impedance transformers are, by far, the most difficult to fabricate. They not only require higher reactances for sufficient isolation (thus more turns), but also windings with higher characteristic impedances. The result is that they are much larger than low-impedance transformers, even though their power capabilities are the same. The size of the core (rod or toroid) is related to the magnitude of the characteristic impedance of the transmission line and the number of turns, and not the power level, since very little flux enters the core.

Sec 8.2   Guanella Transformers

Guanella has shown, in his classic 1944 paper, that by connecting three or more basic building blocks of Fig 1-1 (Chapter 1) in parallel-series arrangements, impedance ratios of 1:9, 1:16...1:n^2 (where n is a whole number) are possible (ref 1).[1] But these transformers are basically bilateral baluns. Either side can have a grounded terminal, resulting in a step-up or step-down balun. If both sides are grounded, as in the unbalanced-to-unbalanced case, the low-frequency performance can be seriously affected. This is especially true in the 1:4 case when both building blocks are wound on the same core (see Chapter 2, Sec 2.3). Extra isolation in the form of a series 1:1 balun or the use of separate cores (with more turns) is found necessary in order to maintain the same low-frequency performance as when operating as a balun. Examples of these were shown in Chapter 6 with 1:4 ununs. The 1:9 and 1:16 cases, which the author has found to require separate cores, also improve their low-frequency responses considerably by the use of series 1:1 baluns. This is especially true at high impedance levels, where the windings, and hence cores, are large. In general, with separate cores for each winding and without a series 1:1 balun, the lowest frequency Amateur

---

[1]Each reference in this chapter can be found in Chapter 15.

Radio band (160 meters) is lost when operating the Guanella transformer as a high transformation ratio (at high impedance levels) unun.

Fig 8-1 shows schematics of the high- and low-frequency models of the Guanella 1:9 transformer. The transformer (Fig 8-1A) is connected in parallel on the left side and in series on the right side, resulting in an output voltage of $3V_1$, and hence an impedance ratio, ρ, of 1:9. Since each transmission line sees, on the right, one-third of R_L, theory predicts an optimum characteristic impedance of $R_L / 3$. In practice, when using low-impedance coaxial cable, the best high-frequency results are obtained with characteristic impedances of about 90% of theory. Practice has also shown that 1:1 baluns, connected in series with Guanella baluns for unun operation, need more reactance (and hence more turns) than the popular 1:1 balun used for isolating Yagi beams and half-wavelength dipoles from coaxial cables.

Like the Guanella 1:4 transformer, the best low-frequency performance occurs when the transformer is connected as a bilateral, 1:9 balun with a floating load, that is, either terminal 1,5,9 or terminal 2 in Fig 8-1A is grounded. But the more interesting case is when the transformer performs as a 1:9 unun (both terminals 1,5,9 and 2 are grounded). In this configuration (with optimized transmission lines), the top transmission line in Fig 8-1A has a gradient of $2V_1$ across its input and output terminals, the middle transmission line has V_1 across its terminals, and the bottom transmission line, zero voltage. Thus, the optimum choice in toroids (or beads) would be ferrites for the top transmission line with a permeability twice that of the middle transmission line. As in the 1:4 case, the bottom transmission line requires no longitudinal reactance, and hence no core or beads. Rod core transformers don't enjoy this flexibility in permeability.

By connecting a 1:1 balun in series at the low-impedance side for a step-up unun, or at the high-impedance side for a step-down unun, and removing the ground at terminal 1,5,9 (and using three cores), an improvement of about a factor of two can be realized in the low-frequency response.

Another interesting case is when the output voltage is balanced to ground, that is, grounds are both on terminals 1,5,9 and 13. In this configuration, the bottom transmission line in Fig 8-1A has a negative gradient of $-3/2V_1$, the middle transmission line of $+V_1$ and the top transmission line of $+1/2V_1$. For best low-frequency performance in this

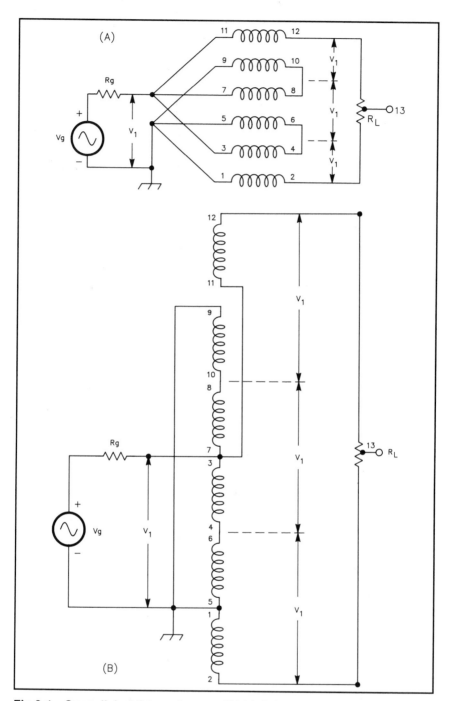

Fig 8-1—Guanella's 1:9 transformer: (A) high-frequency model and (B) low-frequency model. It is assumed that $Z_0 = R_L / 3$, and therefore the output voltages of the transmission lines are each equal to V_1.

case, three cores should be used. Further, the core for the bottom transmission line should have an appropriately higher permeability.

Transformers with ratios of 1:16 and 1:25 can also be designed to perform over wide bandwidths because of the modular nature of Guanella's technique. For example, matching 50 Ω to 800 Ω requires an optimum characteristic impedance of 200 Ω for the transmission lines. This is about the upper limit that can be obtained in power applications with toroids having outside diameters of about 2.5 to 3 inches.

Matching 50 Ω, unbalanced, to 600 or 1000 Ω, unbalanced, requires impedance ratios of 1:12 and 1:20, respectively. These can be obtained by using fractional ratio baluns in series with Guanella baluns. The 50:600-Ω unun can be realized with a 1:1.33 step-up balun in series with a 1:9 balun, or a 1.33:1 step-down balun in series with a 1:16 balun. The 50:1000-Ω transformation can be accomplished by using a 1.25:1 step-down balun in series with a 1:25 balun.

When designing high-power unbalanced-to-unbalanced transformers with ratios greater than 1:4, and working at high impedance levels, one finds the task to be much more difficult. Not only do these transformers require higher reactances (and hence more turns), but the higher characteristic impedances, which are needed in order to achieve adequate high-frequency response, just compound the problem. Experiments by the author have shown that characteristic impedances in the range of 150 to 200 Ω (Fig 5-2), together with Guanella's modular approach of parallel-series connections, makes possible transformers capable of matching 50 Ω to impedances as high as 1000 Ω with good bandwidths and efficiencies. Examples of high-ratio, unbalanced-to-unbalanced transformers are described in the following subsections. Many of these transformers use components described in more detail in Chapters 7 and 9. Of special interest is the fractional-ratio balun.

Sec 8.2.1   5.56:50-Ω Ununs

Fig 8-2 shows a photograph of a low-impedance, 1:9, unbalanced-to-unbalanced transformer designed to match 50-Ω coaxial cable to an unbalanced impedance of 5.56 Ω. The schematic is shown in Fig 8-3. A 1:1 Guanella balun is added, in series, on the high-impedance side in order to improve the low-frequency performance. The 1:1 balun has 10 turns of no. 14 wire on a 1½-inch OD, 4C4 toroid (μ = 125). One of the wires has two layers of Scotch no. 92 tape. This extra separation

Fig 8-2—A Guanella 1:9, unbalanced-to-unbalanced transformer designed to match 5.56 Ω to 50 Ω.

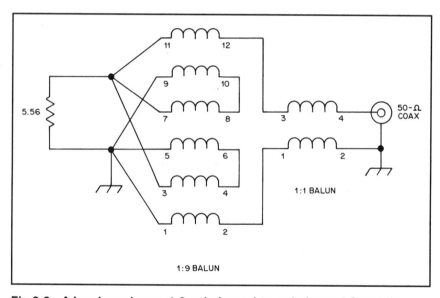

Fig 8-3—A low-impedance, 1:9 unbalanced-to-unbalanced Guanella transformer.

between the wires results in a characteristic impedance of 50 Ω. Each transmission line of the 1:9 balun has 7½ turns of low-impedance cable on a ½-inch diameter, 2½-inch long, no. 61 rod (μ = 125). The coaxial cable uses two layers of Scotch no. 92 tape on no. 14 wire for the inner conductor. The outer braid, which is wrapped with Scotch no. 92 tape, is from RG-122/U cable or equivalent (⅛-inch flat braid can also be used if it is opened up). This transformer has a constant impedance ratio from 3.5 MHz to well over 30 MHz. The highest frequency response occurs at about the 6.67:60-Ω level. If no. 12 wire were used instead, the maximum response would occur at the 5.56:50-Ω level. By using a toroid, for the 1:1 balun, with a permeability of 250 or 290, the low-frequency response would cover the 160-meter band. Further, by using 11 turns of low-impedance coaxial cable on 3½-inch long rods, the transformer would cover 160 meters through 10 meters without the series 1:1 balun. The transformer in Fig 8-2 is conservatively rated at 1 kW of continuous power.

This concept can also be extended to a broadband 1:16 unun. In this case, a 1:16 balun would consist of four similar low-impedance cables connected in parallel on the left and in series on the right. In this higher-ratio case, in order to better match 3.125 Ω to 50 Ω, a thicker wire (than no. 14) would have to be used for the inner conductor. A no. 10 wire with 1½ to 2 layers of Scotch no. 92 tape is recommended. The outer braid remains the same as in the 1:9 case.

Sec 8.2.2   50:300-Ω Ununs

Fig 8-4 shows the schematic of a Guanella 50:300-Ω (1:6) unbalanced-to-unbalanced transformer. Several versions of this transformer

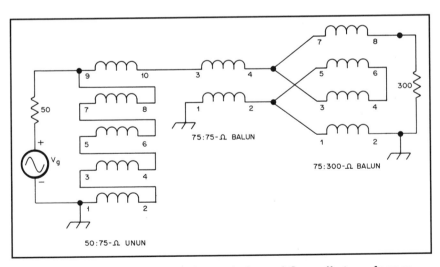

Fig 8-4—A 50:300-Ω unbalanced-to-unbalanced Guanella transformer capable of 1-kW performance from 1.5 MHz to 45 MHz.

have been constructed and found to give broadband responses from 1.5 MHz to 45 MHz, depending upon the various components employed. They all used the quintufilar, 50:75-Ω unun shown in Fig 8-4 and described in detail in Chapter 7. The 75:75-Ω balun, which is described in Chapter 9, has 12 bifilar turns on a 2-inch OD, K5 toroid (μ = 290). Its large reactance, which is some 15 times greater than that of the W2AU balun, was found necessary in order to provide adequate isolation for the 75:300-Ω Guanella balun. This 1:4 (75:300-Ω) balun, which is also described in Chapter 9, has two bifilar windings, of 7 turns each, on a single 2.4-inch OD, no. 64 toroid (μ = 250). As a balun, it performs well from 1.5 MHz to 60 MHz. When used in Fig 8-4, the flat response of the 1:6 unun is from 3.5 MHz to 45 MHz. If this 1:4 balun were replaced with two toroids with 14 bifilar turns each, then the frequency range of the 50:300-Ω unun would be from 1.5 MHz to 45 MHz. A single 2.625-inch OD, K5 toroid (μ = 290) with two bifilar windings of 8 turns each would also provide the same frequency range. As will be shown in Sec 8.2.4 and Chapter 9, the 1:1.56 unun and 1:1 balun can be wound on the same toroid, resulting in a fractional-ratio balun.

Sec 8.2.3   50:450-Ω Ununs

Fig 8-5 is a schematic of a 50:450-Ω unbalanced-to-unbalanced Guanella transformer. It is simpler, in a way, than the transformer of

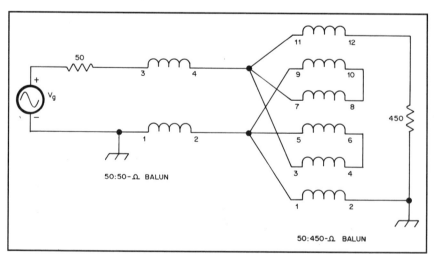

Fig 8-5—A 50:450-Ω unbalanced-to-unbalanced Guanella transformer capable of 1-kW performance from 1.5 MHz to 45 MHz.

Fig 8-4, since only two different series transformers are required; that is, a 1:1 balun in series with a 1:9 balun. With optimum design, this transformer can easily handle 1 kW of continuous power over a frequency range of 1.5 MHz to over 45 MHz. For the 1:1 balun, a 2-inch OD toroid with a permeability of 250 to 290 and with 11 or 12 bifilar turns is recommended (it is described in Chapter 9). The 1:9 (50:450-Ω) Guanella balun is also described in Chapter 9. It uses three 2.625-inch OD, K5 toroids (μ = 290), each with 16 bifilar turns of no. 16 wire (Z_0 = 150 Ω). Without the 1:1 series balun, the frequency range of the 1:9 unun in Fig 8-5 is 3 MHz to over 45 MHz.

Sec 8.2.4   50:600-Ω Ununs

The 50:600-Ω (1:12) unbalanced-to-unbalanced transformer, covering the HF band (3 MHz to 30 MHz), is one of the most difficult to construct. It not only requires three separate series transformers

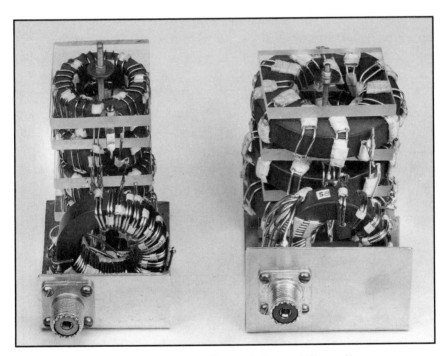

Fig 8-6—Two 50:600-Ω unbalanced-to-unbalanced Guanella transformers designed for operation from 3 MHz to 30 MHz. The transformer on the left is rated at 500 W of continuous power, the transformer on the right, 1 kW.

(unun-balun-balun), but also high characteristic impedance windings in order to achieve good high-frequency performance. The transmission lines, which are then wide-spaced (see Fig 5-2), restrict the number of turns on a practical toroid, and hence limit the low-frequency response. The author was able to design two versions that were capable of flat responses from 3 MHz to 30 MHz. One used a fractional-ratio 1.33:1 step-down balun (two transformers on the same core) in series with a 1:16 step-up balun. It has a 500-W capability and is shown on the left in Fig 8-6. The other transformer used a fractional-ratio 1:1.33 step-up balun in series with a 1:9 step-up balun and is shown on the right in Fig 8-6. It has a 1-kW capability. The parameters for these two transformers are as follows:

A) *Left:*

The schematic for this 500-W unun is shown in Fig 8-7. The 1.33:1 step-down balun uses 4 quintufilar turns of no. 16 wire on a 2-inch OD, K5 toroid (μ = 290). The top winding is tapped one turn from terminal 9. The transformation ratio is 1.4:1. If 5 quintufilar turns were used, and the tap was at 2 turns from terminal 9, the ratio would

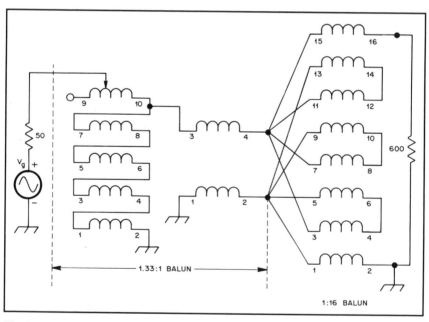

Fig 8-7—A 50:600-Ω unbalanced-to-unbalanced Guanella transformer with a 1.33:1 step-down balun in series with a 1:16 step-up balun.

be 1.32:1 (but little difference would be noted in performance). The 1:1 balun winding (on the same core) has 11 bifilar turns of no. 16 wire. The 1:16 balun has 13 bifilar turns of no. 20 wire on each of the four 1¾-inch OD, K5 toroids ($\mu = 290$). The spacing of the wires is such as to give a characteristic impedance of 150 Ω (see Fig 5-2).

B) *Right:*

The schematic for this 1-kW unun is shown in Fig 8-8. The 1:1.33 step-up balun uses 3 septufilar turns on a 2-inch OD, K5 toroid ($\mu = 290$). The top winding, 13-14, uses no. 14 wire, and the others, no. 16 wire. The 1:1 balun (on the same toroid) has 11 bifilar turns of no. 16 wire, with one wire covered with two layers of Scotch no. 92 tape (for a characteristic impedance of 66.7 Ω). The impedance ratio of this fractional-ratio balun is 1:1.36 (and is close enough to 1:1.33). The 1:9 balun has 11 bifilar turns of no. 18 wire on each of the three 3-inch OD, 4C4 toroids ($\mu = 125$). The spacing is such as to give a characteristic impedance of 200 Ω (see Fig 5-2).

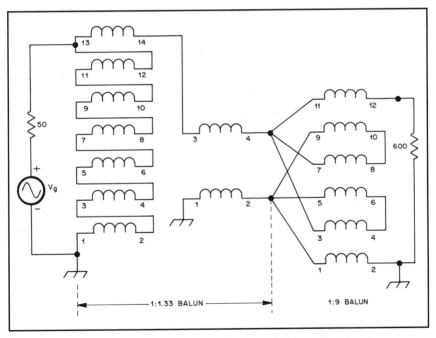

Fig 8-8—A 50:600-Ω unbalanced-to-unbalanced Guanella transformer with a fractional-ratio 1:1.33 step-up balun in series with a 1:9 step-up balun.

Sec 8.2.5   50:800-Ω Ununs

The schematic for the 50:800-Ω unbalanced-to-unbalanced Guanella transformer is shown in Fig 8-9. This transformer is also simpler than the 50:600-Ω ununs of Figs 8-7 and 8-8 since it does not require the extra 1:1.33 unun stage. The 50:50-Ω (1:1) balun has been described above in Fig 8-5. The 50:800-Ω (1:16) balun is a simple extension of the 66.67:600-Ω (1:9) balun described above in Fig 8-8. In this case, four toroids (with similar windings) are used instead of three.

If one is interested in matching 50 Ω unbalanced to 1000 Ω unbalanced (1:20), which is at about the limit of practicality, a suggested configuration is as follows: (A) a 1.25:1 step-down unun in series with a 1:1 (40:40-Ω) balun, in series with a 1:25 (40:1000-Ω) step-up balun; and (B) the 1:25 balun with five toroids having windings similar to those of Fig 8-8.

Sec 8.3   Ruthroff-type Transformers

Another method of obtaining impedance ratios greater than 1:4 in unbalanced-to-unbalanced transformers is an extension of Ruthroff's technique of adding a direct voltage to a voltage which has traversed a coiled transmission line (in this case, several coiled transmission lines). Hence the name—Ruthroff-type transformer. Kraus and Allen, using this technique with a shortened third winding, had previously reported on their transformer yielding a 1:6 ratio (ref 16). The configuration used by the author to obtain ratios as high as 1:9 is shown in Fig 8-10. The device has two input ports, A and B, and three output ports, C, D and E. By using B and D, a 1:2.25 ratio is achieved. Using A and D results in a 1:9 ratio. A and C provide a 1:4 ratio. By using A and tapped-port E, ratios from 1:4 to 1:9 are possible. By using B and E, ratios from 1:1 to 1:2.25 are obtained. Ratios less than 1:4 using trifilar windings are covered in Chapter 7. This chapter deals only with ratios greater than 1:4.

By connecting an input voltage V_1 to terminal A, the voltage at D is $3V_1$ and a 1:9 impedance ratio results. The high-frequency response is less than that of the 1:4 ratio, and much less than that of the 1:2.25 ratio. This is because the output voltage now consists of a direct voltage. a single delayed voltage and a double delayed voltage (the top V_1 in Fig 8-10). The low-frequency response is determined in the same manner as that of the 1:4 unun. For example, the reactance of the lower winding in Fig 8-10 should be much greater than the impedance of the

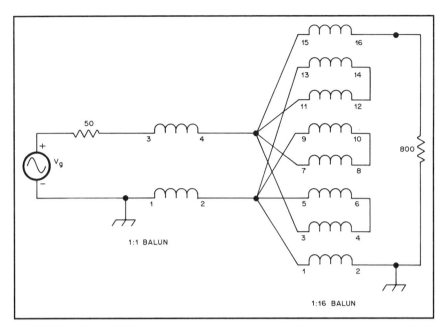

Fig 8-9—A 50:800-Ω unbalanced-to-unbalanced Guanella transformer.

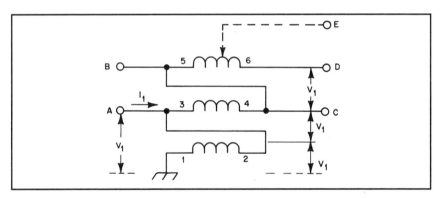

Fig 8-10—A schematic diagram of a trifilar transformer for obtaining impedance ratios up to 1:9.

signal generator. Or conversely, the reactance of the three windings, which will be in series-aiding, should be much greater than the load impedance. With the 1:9 connection, A to D, two-thirds of the input current I_1 flows in the bottom winding and one-third in the top two windings. As can be seen from the schematic, these currents cancel out the flux in the core and the high efficiency of a true transmission line transformer can be achieved.

By varying the tap on the upper winding of Fig 8-10, the connection between terminals A and E yields ratios from 1:4 to 1:9. For a rod transformer, the impedance ratio, ρ, becomes:

$$\rho = (V_0 / V_1)^2 = (2 + l / L)^2 \qquad \text{(Eq 8-1)}$$

where

    l = the length from terminal 5
    L = the length of the winding

For a toroidal transformer, Eq 8-1 becomes:

$$\rho = (2 + n / N)^2 \qquad \text{(Eq 8-2)}$$

where

    n = the number of integral turns from terminal 5
    N = the total number of trifilar turns.

Fig 8-11 shows two tapped trifilar transformers using Q1 toroids of 1¼-inch OD (μ = 125). The transformer on the left consists of seven trifilar turns of ⅛-inch stripline insulated with one layer of Scotch no. 92 tape. The upper winding (in Fig 8-10) is tapped at n = 3, 4 and 5 turns from terminal 5. Fig 8-12 shows the performance of this transformer at the various ratios when terminated in a 50-Ω load. With the maximum loss level set at 0.4 dB, which is equivalent to an SWR of 2:1 (that is, 10% of the power is reflected because of a mismatch), the upper frequency cutoff for all outputs exceeds 30 MHz. The characteristic impedances of the trifilar windings for all ratios, except at 1:4, are optimized for a termination of 50 Ω. The 1:4 ratio response is optimum when R_L = 30 Ω.

Fig 8-13 shows the performance curves of the transformer pictured on the right in Fig 8-11. This device uses seven trifilar turns of no. 14 wire with taps on the upper winding at n = 3 and 5 turns from terminal 5. The impedance ratios, using the A port for the input, are 1:4, 1:5.9, 1:7.37 and 1:9. With R_L = 50 Ω, it can be seen in Fig 8-13 that the 0.4 dB cutoff limit is about 15 MHz for ratios greater than 1:4. With the 1:4 ratio, the upper cutoff is 50 MHz. This shows that for a trifilar winding with no. 14 wire, and with the top winding floating in the 1:4 configuration, a near-optimum condition exists for a 50-Ω load. When the load impedance is 100 Ω or greater, a higher frequency response is obtained

Fig 8-11—Two tapped trifilar transformers designed to give impedance ratios from 1:4 to 1:9. The transformer on the left uses ⅛-inch stripline and the transformer on the right uses no. 14 wire.

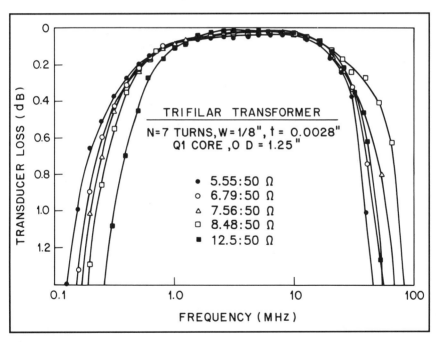

Fig 8-12—Loss v frequency at the 50-Ω level for the tapped trifilar transformer using stripline.

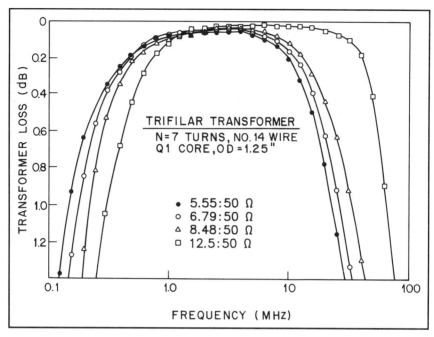

Fig 8-13—Loss v frequency at the 50-Ω level for the tapped trifilar transformer using no. 14 wire. Note the excellent performance at the 12.5:50-Ω level when the third winding is left floating.

at the higher ratios. Fig 8-14 shows the improved performance at the 1:8 ratio of a trifilar transformer using no. 14 wire and loads of 120, 160 and 200 Ω.

Earlier experiments by the author on rod transformers yielded similar results. Fig 8-15 shows the performance of a trifilar transformer with no. 14 wire windings, 22 inches in length, and tightly wound on a Q1 rod ½-inch in diameter and 4 inches long. The characteristic impedance of the windings using the 1:9 connection was 30 Ω, while the characteristic impedance using the 1:4 connection was only 16 Ω (because of the floating third wire). With the 1:9 connection, better high-frequency performance prevails at the 10:90-Ω level. Good high-frequency performance is also obtained at the 9:36-Ω level for the 1:4 ratio because of the influence of the floating third wire. The rod transformer is pictured in the center of Fig 8-16. The two outer transformers in this photo were employed for studies at higher impedance levels using windings with turns spaced by thicker insulation.

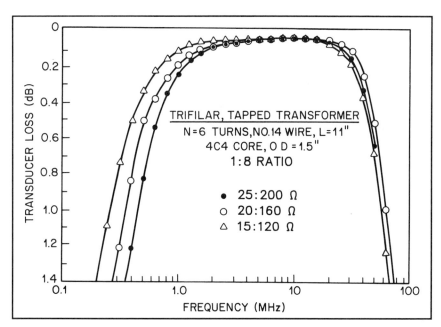

Fig 8-14—Loss v frequency for the tapped trifilar transformer using no. 14 wire at the 1:8 ratio and at three impedance levels.

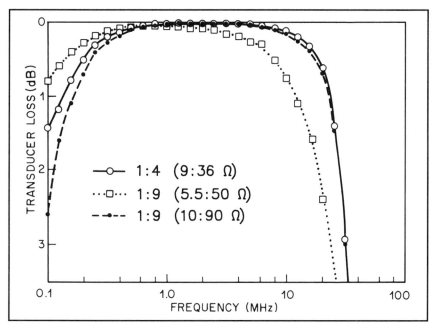

Fig 8-15—Loss v frequency for a trifilar winding on a ½-inch diameter rod yielding impedance ratios of 1:4 and 1:9. The 1:9 data was obtained for two different impedance levels.

Unbalanced-to-Unbalanced Transformer Designs with Impedance Ratios Greater Than 1:4   8-17

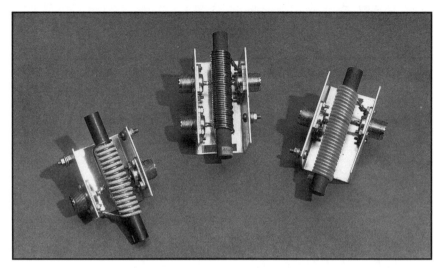

Fig 8-16—Three rod transformers. The measurements shown in Fig 8-15 are for the center transformer.

Sec 8.3.1   5.56:50-Ω Ununs

Figs 8-13 and 8-15 showed that a 1:9 ratio, at the 5.56:50-Ω impedance level, does not result in good high-frequency performance when using no. 14 wire in the schematic of Fig 8-10. These transformers work better at about twice this impedance level. Fig 8-15 also shows that 12 trifilar turns on a ½-inch diameter rod are more than necessary in order to include 1.5 MHz at the low-frequency end. In order to improve the high-frequency performance of these transformers at the 5.56:50-Ω level, the schematic of Fig 8-10 was rearranged as shown in Fig 8-17. By transposing the bottom winding in Fig 8-10 to the middle winding, the characteristic impedance was lowered considerably. By selecting the optimum number of turns, and by using the circuit of Fig 8-17, transformers can be made to adequately cover the 1.5- to 30-MHz range, even though their optimum performances occur at

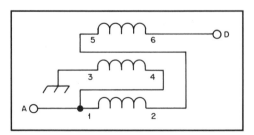

Fig 8-17—A schematic diagram of a 1:9 trifilar transformer with winding 3-4 of Fig 8-10 transposed in order to obtain improved high-frequency performance at the 5.56:50-Ω level.

nearly twice this impedance level. At these higher impedance levels, the high-frequency performances extend well beyond 45 MHz. Fig 8-18 shows four transformers, which use the schematic of Fig 8-17. Their parameters are as follows:

A) *Lower left:*
   5 trifilar turns of no. 14 wire on a 1½-inch OD, 4C4 toroid (μ = 125). The power rating is 1 kW of continuous power.

B) *Upper left:*
   4 trifilar turns on a 1¾-inch OD, K5 toroid (μ = 290). The center winding in Fig 8-17 is no. 14 wire. The other two are no. 16 wire. The power rating is 500 W of continuous power.

C) *Upper right:*
   7 trifilar turns on a ⅜-inch diameter no. 61 rod (μ = 125). The center winding in Fig 8-17 is no. 14 wire. The other two are no. 16 wire. The power rating is 500 W of continuous power.

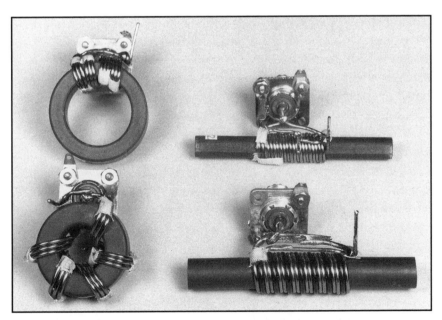

Fig 8-18—Four 1:9 ratio ununs using the schematic of Fig 8-17 in order to obtain operation from 1.5 MHz to 30 MHz at the 5.56:50-Ω level.

D) *Lower right:*

7 trifilar turns on a ½-inch diameter no. 61 rod (μ = 125). The center winding in Fig 8-17 is no. 12 wire. The other two are no. 14 wire. The power rating is 1 kW of continuous power.

If all the transformers in Fig 8-18 used no. 12 wire for the center winding in Fig 8-17 and no. 14 wire for the outer two, they would all have the same power rating of 1 kW of continuous power. As has been noted before, since so little flux enters the core, the power rating is actually determined by the ability of the windings to handle the current, and not by the size of the core.

Sec 8.3.2   50:450-Ω Ununs

As mentioned above, the trifilar transformers of Figs 8-13 and 8-15, which use the schematic of Fig 8-10, are best suited to match at impedance levels considerably higher than 5.56:50-Ω. But going up to a much higher impedance level, like trying to match 50 Ω to 450 Ω, presents a most difficult task for the Ruthroff-type transformer. Not only are many more turns required at this impedance level, but characteristic impedances in excess of 200 Ω only compound the problem. Nevertheless, transformers matching 50 Ω to 450 Ω, having a relatively constant 1:9 impedance ratio from 1.5 MHz to 10 MHz, are rather easy to construct and should have some practical uses. One example is a matching transformer for a Beverage antenna. Two Beverage antenna transformers are shown in Fig 8-19, together with a 1-kW power transformer. Their parameters are as follows:

A) *Left:*

10 trifilar turns of no. 22 hook-up wire on a 1½-inch OD 4C4 toroid (μ = 125). A tap is at 4 turns from terminal 6 in Fig 8-10. Three ratios are available: 1:4, 1:6 and 1:9. This transformer matches 50 Ω to 200, 300 or 450 Ω from 1.5 MHz to 10 MHz. The power rating is 100 W of continuous power.

B) *Center:*

12 trifilar turns on a 3-inch OD, 4C4 toroid (μ = 125). The bottom winding in Fig 8-10 uses no. 14 wire, and the other two, no. 16 wire. All three windings are covered with Teflon tubing (wall thickness about 17 mils). As with the other two smaller transformers, the 1:9 ratio is

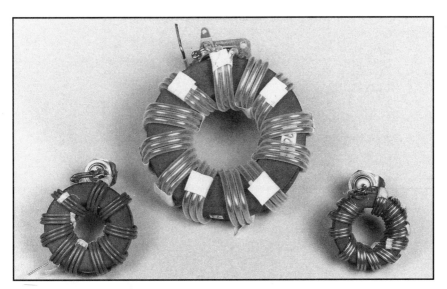

Fig 8-19—Three 1:9 ratio ununs designed to match 50 Ω to 450 Ω from 1.5 MHz to 10 MHz. The two smaller transformers are designed to match 50-Ω coaxial cable to Beverage antennas.

constant from 1.5 MHz to 10 MHz. The power rating is at least 1 kW of continuous power.

C) *Right:*

13 trifilar turns of no. 22 hook-up wire on a 1¼-inch OD, Q1 toroid (μ = 125). Two ratios, 1:4 and 1:9, are available. These ratios are constant when matching 50 Ω to 200 Ω or to 450 Ω from 1.5 MHz to 10 MHz. The power rating is 100 W of continuous power.

Sec 8.3.3   3.125:50-Ω Ununs

As mentioned in the introduction, Chapter 4 presented a 1:16 (3.125:50-Ω) unun using two stripline Ruthroff transformers in series and Chapter 9 presents a 1:12 (50:600-Ω) balun, also using two Ruthroff transformers in series. The final part in this section presents three other Ruthroff-type transformers yielding 1:16 ratios, at low impedances, with only single cores. One uses two coaxial-cable 1:4 ununs in series, as shown in Fig 8-20A, and the other two use quadrifilar windings, as shown in Fig 8-20B. Fig 8-21 shows a photograph of the three transformers. The parameters for the three transformers are as follows:

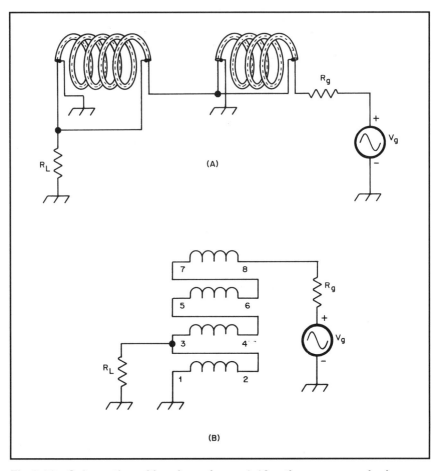

Fig 8-20—Schematics of low-impedance 1:16 ratio ununs employing single cores: (A) two coaxial cable 1:4 ununs in series and (B) a quadrifilar connection.

Fig 8-21—Three low-impedance 1:16 ununs employing single cores.

A) *Left:*

The low-impedance unun (on the left in Fig 8-20A) has 4 turns of coaxial cable. The inner conductor is no. 10 wire with about 1½ layers of Scotch no. 92 tape (because it is wrapped edgewise; see Chapter 13). The outer braid, from RG-122/U, is wrapped with Scotch no. 92 tape. The characteristic impedance is 9 Ω. The higher-impedance unun (on the right in Fig 8-20A) has 7 turns of coaxial cable. The inner conductor, of no. 16 wire, has 2 layers of Scotch no. 92 tape. The outer braid, from RG-122/U, is also wrapped with Scotch no. 92 tape. The characteristic impedance is 14 Ω. The toroid is a 1½-inch OD, 4C4 core ($\mu = 125$). The transformer covers 1.5 MHz to 30 MHz when matching 3.125 Ω to 50 Ω. The highest frequency response occurs when matching 3.875 Ω to 62 Ω. The power rating is 1 kW of continuous power.

B) *Center:*

5 quadrifilar turns on a ½-inch diameter, no. 61 rod ($\mu = 125$). The bottom winding in Fig 8-20B uses no. 14 wire, and the other three windings use no. 16 wire. At the 3.125:50-Ω level, the impedance ratio is constant from 1.5 MHz to 10 MHz. The highest frequency response occurs at the 6.25:100-Ω level. At this impedance level, the impedance ratio is constant up to 25 MHz. The power rating is 500 W of continuous power.

C) *Right:*

4 quadrifilar turns on a 1¼-inch OD, K5 toroid ($\mu = 290$). The bottom winding in Fig 8-20B uses no. 14 wire, and the other three use no. 16 wire. At the 3.125:50-Ω level, the impedance ratio is constant from 1.5 MHz to 15 MHz. The optimum performance also occurs at the 6.25:100-Ω level. The power rating is 250 W of continuous power.

Sec 8.4 Ruthroff-Guanella Transformers

Another technique which is presented in the literature for a 1:9 unbalanced-to-unbalanced transformer is shown in Fig 8-22 (ref 29). This is an extension of Guanella's method on connecting transformers in a parallel-series arrangement. In Fig 8-22, the top two windings form a 1:1 balun transformer. The bottom two windings are connected to form a 1:4 Ruthroff unun. The transformers are in parallel on the left side and in series on the right. Since the outputs of the transformers are isolated from the inputs in the passband because of the reactance of the coiled windings, the output voltage is $V_O = 3V_1$. This results in a 1:9 impedance ratio when the output is taken from terminal B to ground. If the output is taken from terminal C to ground, a 1:4 ratio is obtained. Fig 8-23 is a photograph of two 1:9 Ruthroff-Guanella transformers together with a Ruthroff-type (with a single core) transformer which was used as a vehicle for comparison. The parameters for these transformers are as follows:

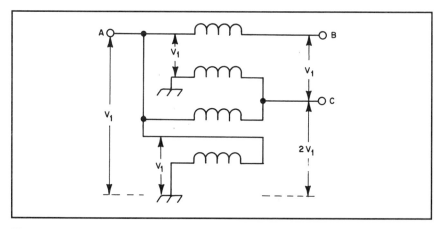

Fig 8-22—A schematic of a 1:9 and 1:4 Ruthroff-Guanella transformer using two basic building blocks.

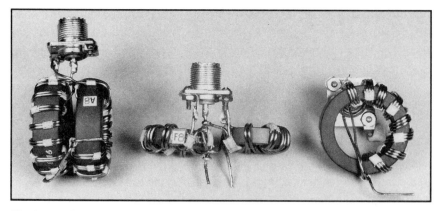

Fig 8-23—Three 1:9 ratio ununs. The transformer on the right is a Ruthroff-type unun (Fig 8-10) with a bandwidth of 1.5 MHz to 15 MHz. The other two are Ruthroff-Guanella ununs (Fig 8-22) with bandwidths of 1.5 MHz to 45 MHz.

A) *Left:*

The top two windings in Fig 8-22 are 16 bifilar turns of no. 16 wire. The bottom two windings are 8 bifilar turns of no. 16 wire. The cores are 1¾-inch OD, K5 toroids (μ = 290). The useful impedance-level range is 13.33:120-Ω to 20:180-Ω, where the impedance ratio is constant from 1.5 MHz to 30 MHz. When matching 15.56:140-Ω, which is the optimum impedance level, the ratio is constant from 1.5 MHz to 45 MHz. The power rating is 1 kW of continuous power.

B) *Center:*

The top two windings in Fig 8-22 are 7 bifilar turns of no. 16 wire. The bottom two windings are 10 bifilar turns of no. 16 wire. The cores are 1¼-inch OD, K5 toroids (μ = 290). Because of the rather short lengths of transmission lines, this transformer has a wide impedance-level capability. When matching 8.33 Ω to 75 Ω, the impedance ratio is constant from 1.5 MHz to 30 MHz. When matching 22.22 Ω to 200 Ω (which is the optimum), the impedance ratio is constant from 5 MHz to 45 MHz. In between these two impedance levels, the transformer can be said to be useful from 3.5 MHz to 30 MHz. The power rating is 500 W of continuous power.

C) *Right:*

This Ruthroff-type transformer has 8 trifilar turns of no. 16 wire on a 1¾-inch OD, K5 toroid (μ = 290). It is connected as shown in Fig 8-10. In comparing with transformer (B), above, the useful frequency range is found to be about one-half, that is, only 1.5 MHz to 15 MHz. There is little doubt that the Ruthroff-Guanella transformer is superior to the Ruthroff-type transformer in this range of impedance levels.

From the results on these three transformers and from other Ruthroff-Guanella transformers, the following comments are offered:

1) The Ruthroff-Guanella transformer requires two cores for best operation.

2) When using low-impedance coaxial cables (for example, when matching 5.56 Ω to 50 Ω), satisfactory operation was obtained from 1.5 MHz to only 25 MHz. Beyond 25 MHz, serious resonances occurred.

3) The Ruthroff-Guanella transformer was difficult to work with at impedances greater than 300 Ω.

4) The top two bifilar windings determine, in large measure, the low-frequency response. It is recommended that they have twice the number of bifilar turns as the bottom two windings. With fewer turns than this, the high-frequency performance improves, but at the expense of the low-frequency performance.

5) The regular Guanella transformer with its parallel-series connection of transmission lines is by far the best transformer.

Sec 8.5   Coaxial Cable Transformers—Ruthroff-Type

As has been shown throughout this book, many wire versions of transmission line transformers can be converted to coaxial cable transformers. The resulting advantages are higher current and voltage capabilities and less parasitic coupling between adjacent turns. Further, they also lend themselves, quite readily, to multiport operation; that is, they can possess more than one broadband impedance ratio. They can also be tapped, yielding fractional ratios. Fig 8-24 shows a schematic diagram for a trifilar configuration yielding ratios of 1:2.25 with terminals A and B, 1:4 with terminals B and C, and 1:9 with terminals A and C. Fig 8-25 shows a trifilar, coaxial cable transformer using two sections of RG-58/U. The outer covering was removed for ease of winding. As shown in Fig 8-24, the outer braids are at the same potential and therefore can be in contact with each other. Fig 8-26 shows the performance with the 1:9 ratio for three different levels of output impedance:

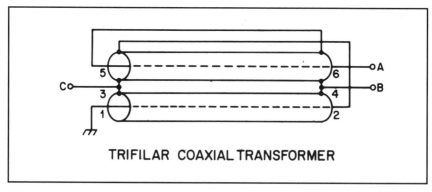

TRIFILAR COAXIAL TRANSFORMER

Fig 8-24—The schematic diagram of a trifilar transformer that uses two sections of coaxial cable, yielding impedance ratios of 1:2.25, 1:4 and 1:9.

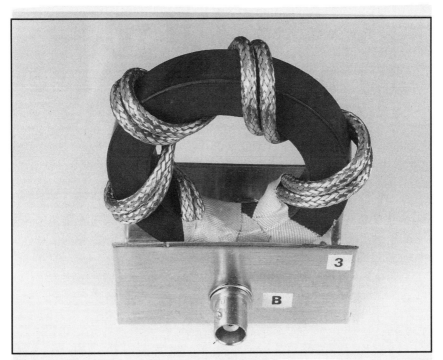

Fig 8-25—The trifilar transformer using two sections of RG-58/U coaxial cable.

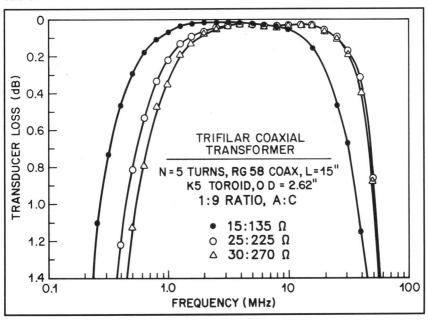

Fig 8-26—Loss v frequency of the transformer of Fig 8-25 as a function of impedance levels at the 1:9 connection.

135, 225 and 270 Ω. At the high-frequency end, the 0.4-dB points are at 40 MHz for the 225- and 270-Ω loads. Surprisingly, between 2 and 25 MHz, the losses are less than 0.1 dB at these impedance levels. This is considerably less than wire transformers using this rather high permeability K5 material (μ = 290), shown in Chapter 11. The 1:2.25 performance shown in Fig 8-27 is equally interesting. The upper cutoff frequencies approach 80 MHz with 110- and 135-Ω loads. Fig 8-28 shows the performance for the 1:4 connection. These results show the influence of the trifilar connection. Optimum performance now occurs at the 18.75:75-Ω level instead of the 25:100-Ω level, as shown in Fig 3-6. Fig 8-28 also shows a greater slope in the loss with frequency than for the other two ratios. This slope is typical of wire transformers using K5 material.

Fig 8-29 shows two low-impedance versions of the trifilar, coaxial cable transformer. Their parameters are as follows:

A) *Left:*

5 trifilar turns (with two coaxial cables) on a 1½-inch OD, 4C4 toroid (μ = 125). The coaxial cable consists of no. 14 wire with two

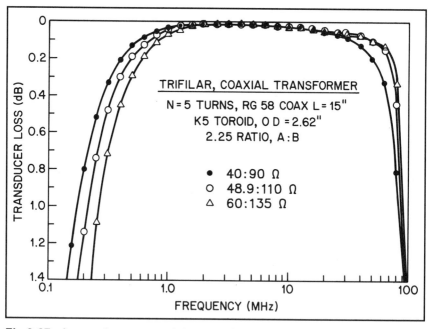

Fig 8-27—Loss v frequency of the transformer of Fig 8-25 as a function of impedance levels at the 1:2.25 connection.

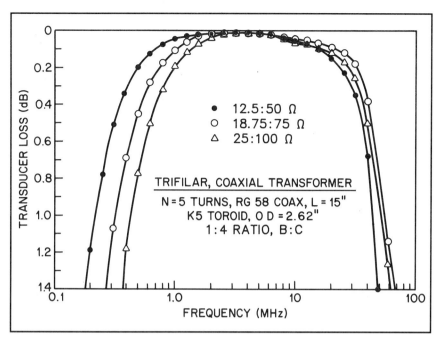

Fig 8-28—Loss v frequency of the transformer of Fig 8-25 as a function of impedance levels at the 1:4 connection.

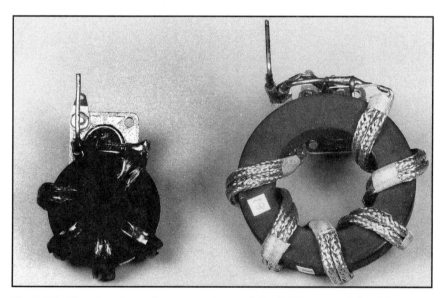

Fig 8-29—Two low-impedance versions of the trifilar, coaxial cable transformer.

layers of Scotch no. 92 tape for the inner conductor, and braid from RG-122/U (or equivalent), wrapped with Scotch no. 92 tape, for the outer conductor. The characteristic impedance, Z_0, is 14 Ω. With the 1:9 ratio, the response at the 5.56:50-Ω level is essentially flat from 1.5 MHz to 30 MHz. At the 11.11:100-Ω level, which is optimum, it is flat from 1.5 MHz to 50 MHz. Very wideband responses were also observed when matching 22.22 Ω to 50 Ω with the 1:2.25 ratio and 12.5 Ω to 50 Ω with the 1:4 ratio.

B) *Right:*

6 trifilar turns (with two coaxial cables) on a 2.4-inch OD, no. 61 toroid ($\mu = 125$). The coaxial cable is the same as above except that the outer braid is not wrapped tightly with Scotch no. 92 tape. The characteristic impedance in this case is 17 Ω. Although the responses at the different impedance ratios and impedance levels, as noted above in (B), are quite adequate, the performance by the smaller transformer (A) is better because of its lower characteristic cable and its shorter lengths of transmission line. The larger transformer (B) favors impedance levels some 25% higher than the smaller transformer (A).

Chapter 9

Baluns

Sec 9.1   Introduction

The balun (*bal*anced-to-*un*balanced) transformer is a subset of the general class known as transmission line transformers. This class differs greatly from the conventional class of transformers where energy is transmitted from input to output by flux linkages. Here, energy is transmitted by a transmission line mode, and the conventional current flow is prevented by the choking action of the coiled transmission lines. This conventional current flow can lead to high flux densities and saturation when the balun is improperly designed or the impedance, as seen at the output, is much greater than expected. The balun, as well as other forms of the transmission line transformer, does not have a primary winding or a secondary winding as such. Its operation in the passband, where only transmission line currents flow, is analyzed purely by transmission line theory. The objectives then are: (A) to have sufficient reactance in the coiled transmission lines to prevent the unwanted conventional current, (B) to select the proper characteristic impedance in order to optimize the high-frequency response and (C) to minimize the parasitics which eventually reduce the inductive reactance of the coiled transmission lines at high frequencies to unacceptable values.

To date, practically all of the 1:1 and 1:4 balun designs, excluding baluns using ferrite beads or ½-λ and ¼-λ sections of transmission line, have used the schematics of Ruthroff which were presented in his classic paper in 1959 (ref 9).[1] But the earlier paper by Guanella in 1944 also presented schematics for the same baluns (ref 1). For reasons unknown to the author, Guanella's baluns were not favored. After examining many forms of the balun over the past two years, and from feedback from the first edition of this book and discussions with many colleagues, it became apparent to the author that Guanella's approach is the preferred one.

1Each reference in this chapter can be found in Chapter 15.

Therefore, this chapter is completely revised from the first edition. Designs are presented which are different than the designs which have been most popular over the past three decades. They lead to a host of new applications. These designs include baluns feeding Yagi and quad antennas directly from 50-Ω coaxial cable and very broadband baluns capable of matching from 50 Ω to 3.125 Ω at the low-impedance end and 50 Ω to 1000 Ω at the high-impedance end.

Sec 9.2   The 1:1 Balun

The 1:1 balun is well known to radio amateurs and antenna professionals, since it is widely used to match coaxial cables to dipole antennas and to Yagi beams that incorporate matching networks which raise the input impedance to that of the cable. The purpose of the balun is to minimize RF currents on the outer shield of the coaxial cable which would otherwise distort radiation patterns (particularly the front-to-back ratio of Yagi beams) and also cause problems because of RF penetration into the operating room. The balun accomplishes this by its balanced feed and choking action at the low-impedance point (and hence high-current point) on an induced antenna current due to antenna asymmetry. Many successful forms of the 1:1 balun have been used. They include: (A) the bazooka, which uses $\frac{1}{4}$-λ decoupling stubs, (B) some 10 turns of the coaxial line with a diameter of 6 to 8 inches, (C) ferrite beads over the coaxial line, and (D) ferrite-core or air-core Ruthroff designs.

The most popular form of the 1:1 balun has probably been the Ruthroff design which is shown in Fig 9-1. Fig 9-1A is the schematic for the toroidal version and Fig 9-1B for the rod version. The single winding of the toroidal transformer, shown as an extension of the top winding of the coiled transmission line, is usually wound on its own part of the toroid. The low-frequency model for the 1:1 baluns in Fig 9-1 is shown in Fig 9-2.

Fig 9-1 shows a potential gradient of $-V_1$ across the series windings 3-4 and 5-6. Therefore, terminals 4,5 in (A) and (B), which are at the midpoint of the gradient, have a $+V_1 / 2$ potential to ground. Terminals 2 in (A) and (B) then have a $(V_1 / 2 - V_2)$ potential to ground. When the transmission lines (windings 1-2 and 3-4) are terminated in their characteristic impedances ($R_L = Z_0$), then $V_2 = V_1$ and terminals 2 in (A) and (B) have a $-V_1 / 2$ potential to ground. Thus, the output voltage is balanced to ground or to a common potential. The center of

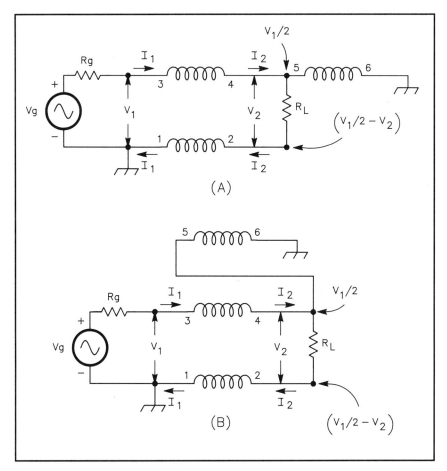

Fig 9-1—High-frequency models of Ruthroff's 1:1 balun: (A) toroidal version and (B) rod version.

R_L can then be connected to terminal 1 since they are now at the same potential.

The third wire, winding 5-6, according to Ruthroff's 1959 paper, is necessary to complete the path for the magnetizing current (ref 9). From recent discussions with colleagues, including Ruthroff, it was agreed that the third wire is not necessary in the

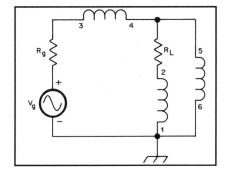

Fig 9-2—Low-frequency model of Ruthroff's 1:1 baluns.

performance of his 1:1 balun in antenna applications. When the reactance of the windings is much greater than R_L (at the lowest frequency of interest), then only transmission line currents flow and the magnetizing current, as such, is irrelevant. In fact, the third wire (winding 5-6), which places terminal 5,6 at a $+V_1 / 2$ potential, can have some negative effects. If R_L is not equal to Z_0, then V_2 is not equal to V_1 and the center of R_L is not at ground (or common) potential. Therefore, the output voltage is not balanced to ground. Further, at very low frequencies (where the reactance of the coiled windings is not much greater than R_L) the shunting effect of winding 5-6 (in Fig 9-2) can cause a conventional current to flow in windings 3-4 and 5-6, creating excessive flux in the core and possible damaging results. However, the third wire has been found to have a beneficial effect on the rod version of Ruthroff's 1:1 balun. Since the windings are generally tight (for example, the W2AU balun), the third wire acts as an electrostatic shield between adjacent bifilar turns and raises the characteristic impedance to reasonable levels.

On the other hand, the Guanella 1:1 balun shown in Fig 9-3, which is nothing more than a coiled bifilar winding, does not have the negative effects of the Ruthroff balun. At very low frequencies, the major currents in the windings are still equal and opposite, and very little flux enters the core. Further, in the passband where the load is isolated from

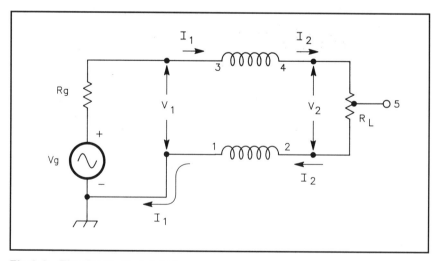

Fig 9-3—The Guanella 1:1 balun.

the input of the 1:1 balun (due to coiling), the center of R_L (terminal 5 in Fig 9-3) is always at ground or common potential.

The input impedances of the 1:1 baluns in Figs 9-1 and 9-3 are the same as that of a terminated transmission line. Assuming sufficient choking such that only transmission line currents flow and neglecting parasitics between adjacent bifilar turns, they are:

$$Z_{in} = Z_0(Z_L + jZ_0\tan\beta l) / (Z_0 + jZ_L\tan\beta l) \qquad \text{(Eq 9-1)}$$

where

Z_0 = the characteristic impedance
Z_L = the load impedance
l = the length of the transmission line
$\beta = 2\pi / \lambda$, where λ = the effective wavelength in the transmission line.

It can be seen from Eq 9-1 that the input impedance can be complex, except when $Z_0 = Z_L$, and is periodic with the variation of βl—the period being π or $l = \lambda / 2$. For short transmission lines, that is, $l < \lambda / 4$, the impedance is less than Z_L if Z_L is greater than Z_0, and greater than Z_L if Z_L is less than Z_0. In other words, the transformation ratio is greater than 1:1 if Z_L is less than Z_0 and less than 1:1 (like 0.5:1) if Z_L is greater than Z_0. This variation in the transformation ratio becomes apparent when the length of the transmission line becomes greater than 0.1 λ.

It should be noted that the characteristic impedance of the 1:1 balun is assumed to be the same as that of the coaxial cable which is connected to its terminals. This is true with the Guanella balun using no. 14, or 16 wire with very little spacing between the wires and ample spacing (at least one-diameter spacing—see Fig 5-2) between adjacent bifilar turns. However, when extra insulation such as Teflon tubing is employed, the characteristic impedance can become two or three times greater than that of the coaxial cable, and the input impedance can differ widely from that of 50-Ω cable, even at reasonably low frequencies.

Fig 9-4 shows three 1:1 baluns, capable of handling 1 kW of continuous power, using ½-inch diameter rods with permeabilities of 125. The parameters for these baluns are:

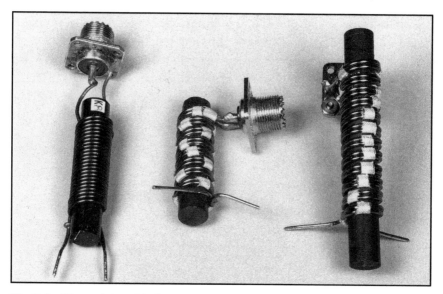

Fig 9-4—Three 1:1 rod baluns.

A) *Left:*

A commercial W2AU (Ruthroff design) balun using 8 trifilar turns of no. 14 wire on a 2½-inch long rod. The characteristic impedance is 43 Ω. The third wire, winding 5-6 in Fig 9-1B, is placed between the other two windings. Without this third wire, which acts as an electrostatic shield, the characteristic impedance of a tightly wound bifilar winding of no. 14 would be 25 Ω. This balun has been widely used on triband (10, 15 and 20-meter) Yagi beams. At much lower frequencies, the performance becomes marginal. It is recommended that this balun not be used below 3.5 MHz.

B) *Center:*

A Guanella balun using 8 bifilar turns of no. 14 wire on a 2⅜-inch long rod. One of the windings is covered with two layers of Scotch no. 92 tape. This added insulation, together with the wire-diameter spacing between adjacent bifilar turns, raises the characteristic impedance to 50 Ω. The low-frequency response of this balun is the same as that of the W2AU balun. If a polyimide-coated wire such as ML or H Imideze wire were to be used, together with one wire having the two layers of Scotch no. 92 tape, the breakdown of this transformer would rival that of coaxial cable. As above, this is a satisfactory balun for triband Yagi beams.

C) *Right:*

A Guanella balun using 12 bifilar turns of no. 14 wire on a 4-inch long rod (a 3-inch long rod would do as well). The treatment of the wire is the same as (B) above. This balun is capable of operating from 1.7 MHz to 30 MHz. If operation is limited to the 40, 80 and 160-meter bands, then 14 bifilar turns are recommended. This would allow for more margin on 160 meters.

Sec 9.2.1 Rod v Toroidal Baluns

Since the 1:1 baluns in Fig 9-4 appear to satisfy most Amateur Radio needs and are the simplest and most inexpensive baluns to construct, then why use toroids for the cores? Here are several answers that can be given to this question:

1) Because the toroidal transformer has a closed magnetic path, and hence much less reluctance than the rod transformer, and because the permeability plays a direct role in the reactance of the coiled windings (rod transformers are independent of permeability—see Fig 2-7), much greater margins can be obtained at both the low- and high-frequency ends of operation. This is achieved by being able to use fewer turns to obtain the desired margins. From Fig 2-5, it can be seen that the low-frequency response with the toroid is better by a factor of 2.5 (which is also a ratio of their inductances) over the rod. Further, higher permeabilities, like 250 to 300, can still be used with good efficiencies. Thus, the overall improvement is greater by a factor of at least 5.

2) The toroid lends itself more readily to the use of thicker wires (like nos. 10 and 12) and coaxial cables. This allows for higher power levels. Since fewer turns are needed with toroids, the spacing between the bifilar turns or the coaxial cable turns can be increased. This lowers the parasitic couplings and increases the high-frequency response.

3) If 1:1 baluns are required in the 75-Ω to 200-Ω range, then the toroidal core is, without doubt, the best choice. The rod core would greatly restrict the useful bandwidth.

4) Many symmetrical forms of the hybrid transformer, as well as Guanella baluns which are employed in unbalanced-to-unbalanced operations, use 1:1 baluns in series with 1:4 (and higher) ratio baluns. The

Fig 9-5—Three Guanella 1:1 toroidal baluns.

extra isolation (from ground) offered by these 1:1 baluns is necessary in order to preserve the low-frequency response. The best isolation is obtained with toroidal core 1:1 baluns.

Fig 9-5 shows three 1:1 Guanella baluns using toroids. Their parameters are as follows:

A) *Left:*

11 bifilar turns of no. 12 H Imideze wire on a 2-inch OD, K5 toroid (μ = 290). Each wire has two layers of Scotch no. 92 tape, resulting in a characteristic impedance of 50 Ω. The power and voltage capabilities are approximately 5 kW of continuous power and 5 kV, respectively. The useful bandwidth extends from 1 MHz to well over 50 MHz. In comparison with the W2AU balun shown in Fig 9-4, an engineering estimate (by the author) would say that this transformer is 5 to 10 times better on power, voltage and bandwidth. This transformer can also be duplicated with no. 14 H Imideze or ML wire, with only one winding having two layers of Scotch no. 92 tape. The power and voltage capability would then be reduced only by a factor of about 2.

B) *Center:*

This transformer, as shown in Fig 9-5, is connected as a 1:4 Ruthroff balun. The following information refers to its connection as a 1:1 Guanella balun. It has 10 turns of RG-58C/U cable on a 2.4-inch

OD, no. 61 toroid (μ = 125). As a 1:1 Guanella balun, the useful bandwidth is from 2 MHz to well over 50 MHz. With a no. 66 toroid (μ = 250) or equivalent, the low-frequency response would extend down to 1 MHz. The power capability is probably greater than 2 kW of continuous power. The voltage breakdown is on the order of 2 kV. The outer jacket was removed in order to make winding easier. Since the inner conductor is rather small, the transformer had to be supported as shown in Fig 9-5. RG-8X cable is not recommended in this application since the insulation, which is foamed polyethylene, would not prevent migration of the inner conductor (and hence shorting) due to its small radius of curvature.

C) *Right:*

This is an example of a 75-Ω, 1:1 Guanella balun. It has 12 bifilar turns of no. 14 wire on a 2-inch OD, K5 toroid (μ = 290). One wire is covered with 17-mil wall Teflon tubing in order to increase the characteristic impedance to 75 Ω (see Fig 5-2). The response is useful from 1.5 MHz to over 50 MHz. The power rating is about 2 kW of continuous power. The voltage breakdown, with H Imideze or ML wire, is in excess of 2 kV. With the help of Fig 5-2, 1:1 baluns can be constructed with characteristic impedances up to 200 Ω.

In Sections 9.2.2 and 9.2.3, models and experimental data are presented on two other (and somewhat controversial) aspects of 1:1 baluns. This information involves their use with shunt-fed dipoles and the comparison between air-core and ferrite-core structures.

Sec 9.2.2   Bifilar v Trifilar Baluns

Most Yagi beam antennas employ a shunt-type feed system in order to raise the input impedance to values near that of coaxial cables. In many cases this results in the center of the driven element being at a common or ground potential. Physically and mathematically, this can be represented by the input impedance of the Yagi beam being grounded at its center-point. This beam antenna, which is balanced to ground, is then matched to an unbalanced coaxial cable by a 1:1 balun. The question then is, how do the Ruthroff and Guanella baluns work in this case? Obviously the trifilar Ruthroff balun performs satisfactorily as witnessed by its overwhelming use. What about the bifilar Guanella balun? In order to understand the problems involved with both of these baluns, the low-frequency models shown in Fig 9-6 were studied, together with measurements obtained on baluns with "floating" loads (like that of dipoles) and grounded, center-tapped loads (like that of Yagi beams). The numbers on the ends of the windings in Fig 9-6 were taken from their high-frequency models, that is, Fig 9-3 for the Guanella model and Fig 9-1 for the Ruthroff model. The low-frequency models in Fig 9-6 assume that the transmission lines formed by windings 1-2 and 3-4 (in both cases) are completely decoupled as far as transmission line operation is concerned, and that the windings are only coupled by flux linkages. There is virtually no energy being transmitted by a transmission line mode in this case. This condition arises when R_L is much greater than the reactances of the individual windings. As can be seen from Fig 9-6, if the frequency is lowered to a point where the reactances approach zero, then the impedance of the Guanella balun approaches $R_L / 2$, and the impedance of the Ruthroff balun approaches zero. But if the reactances of the windings are much greater than R_L (at least 10 times greater), then energy is mainly transmitted to the load by a transmission line mode and the grounding of the center of the load (or at either end) becomes unimportant. In order to prove this point, baluns from Fig 9-4 and Fig 9-5 were measured for their low-frequency responses with the center-tap of R_L grounded (where the problem could arise) with the following results:

1) The rod-type (W2AU) balun on the left in Fig 9-4: The low-frequency response at 3.5 MHz is the same whether the center of R_L is

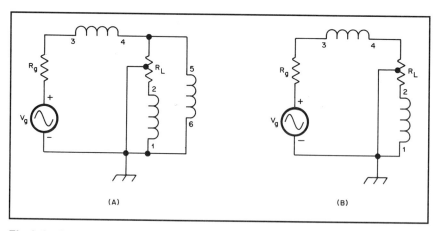

Fig 9-6—Low-frequency models of 1:1 baluns with the center-tap of the load, R_L, grounded: (A) Ruthroff model and (B) Guanella model.

grounded or not. The shunting effect of winding 5-6 dominates at the low end. In any case, grounded or not, it is recommended that this balun only be used above 3.5 MHz.

2) The rod-type Guanella balun in the center in Fig 9-4: The acceptable low-frequency response is 7 MHz with the center of R_L grounded. This balun, which only has 8 bifilar turns (on a rod), should not be used below 7 MHz for fear of high core flux and the consequences thereof. As noted before, without the center of R_L being grounded, the low-frequency limit is 3.5 MHz.

3) The rod-type Guanella balun on the right in Fig 9-4: The acceptable low-frequency response is 3.5 MHz with the center of R_L grounded. Since this balun has 12 bifilar turns, the reactance of the windings is twice that of (2) above, and hence the improvement in the low-frequency response. Without the center of R_L being grounded, the low-frequency limit is 1.7 MHz.

4) The toroidal-type Guanella balun on the left in Fig 9-5: This balun, which has 11 bifilar turns on a 2-inch OD, K5 toroid ($\mu = 290$), has a much greater reactance in its windings than the rod transformers above. Measurements showed that the acceptable low-frequency response with the center of R_L grounded is as low as 1 MHz. This is the balun to use if a shunt-fed Yagi beam were to be used on 160 meters!

Sec 9.2.3   Air-core v Ferrite-core Baluns

Conventional wisdom on baluns uses terms like magnetizing currents, polarity, saturation, core loss and so on. These are terms and concepts related to the conventional transformer, which transmits the energy by flux linkages. This is not the language of the transmission line transformer. When ferrite-core transmission line transformers are designed and used properly (that is, when the reactances of the coiled windings, at the lowest frequency of interest, are at least 10 times greater than the effective termination of the transmission lines) the currents that then flow are mainly transmission line currents. This mode of operation leads to the very wide bandwidths and the exceptional efficiencies (virtually no core loss) that are achievable with these transformers. Therefore, the language of the transmission line transformer is that of RF chokes and transmission lines. In fact, the core losses and the high-frequency responses of 1:1 baluns are primarily determined by the properties of the coiled transmission lines performing as RF chokes.

Air-core baluns do eliminate the core problems when the wrong ferrite is used or when the reactance of the coiled windings is insufficient to suppress the longitudinal currents which create core flux. These currents are the conventional transformer currents and the induced antenna currents on the feed line. With good antenna and feed-line symmetry, and the proper choice of feed-line length (the length to ground of the feed-line and one-half of the dipole should not be an odd multiple of a ¼ λ), the induced currents can become insignificant. But the problem with air-core baluns is the inordinate number of turns required to achieve reactances comparable to ferrite-core baluns. Sec 2.4, in Chapter 2, shows that a winding on a 4-inch long ferrite rod has more than 10 times the reactance of a similar air-core winding. When comparing the reactance of an air-core winding with a similar one on a toroid (see Sec 2.4), the difference is even more dramatic: 27 times more reactance with a toroid of permeability 125. With a toroid of 290 permeability, this amounts to a difference of 62 times. In other words, to equal the reactance provided by a ferrite toroid with a permeability in the range of 250 to 300 (which still provides 98% to 99% efficiency at the 50-Ω level), the air-core balun would require 7 to 8 times more turns. Therefore, to approach the isolation properties of the balun shown on the left in Fig 9-5, an air-core balun would require some 80 bifilar turns. But this is not the whole story. One has to consider the high-frequency properties of an RF choke with 80 turns. Obviously, its self-resonance (which is the limiting factor at the high-frequency end) is much lower in frequency than that of the 11-turn balun of Fig 9-5. In summary, the air-core balun does eliminate potential core problems associated with ferrite-core baluns, but at the expense of bandwidth.

Sec 9.3   The 1:4 Balun

The 1:4 balun, although not as popular as the 1:1 balun, has found considerable use in antenna applications. These include matching folded dipoles to coaxial cables and matching balanced feed lines to unbalanced networks in antenna tuners. Like the 1:1 baluns in Sec 9.2, there are also two forms of the 1:4 balun: (1) the Ruthroff balun, and (2) the Guanella balun. Their high-frequency models are shown in Fig 9-7; Fig 9-8 shows their low-frequency models. But, unlike the 1:1 baluns, both forms of the 1:4 balun have a considerable difference in their performances, depending on whether the load, R_L, is "floating" or

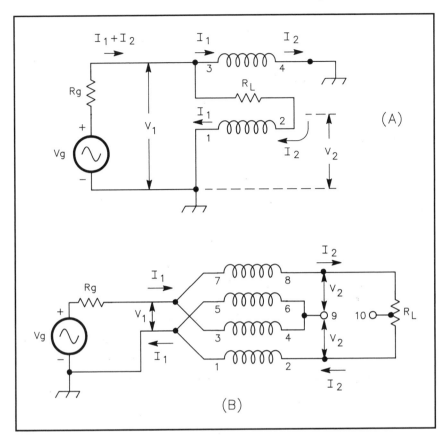

Fig 9-7—High-frequency models of 1:4 baluns: (A) Ruthroff model and (B) Guanella model.

connected to ground (or a common point) at its mid-point. The analyses of these two conditions are as follows:

(A) R_L Floating

The Ruthroff balun (Fig 9-7A) works on the principle that the left side of R_L is at $+V_1$ due to the direct connection to terminal 3 and the right side is at $-V_1$ (for a matched transmission line) because of the negative gradient across both windings (see Sec 1.2). Then $V_{out} = 2V_1$ and the impedance ratio is 1:4. Further, the output is balanced to ground. If the reactances of the windings are much greater than R_L (at the frequency of interest), the currents that then flow are only transmission line currents. The high-frequency performance is the same as that of the

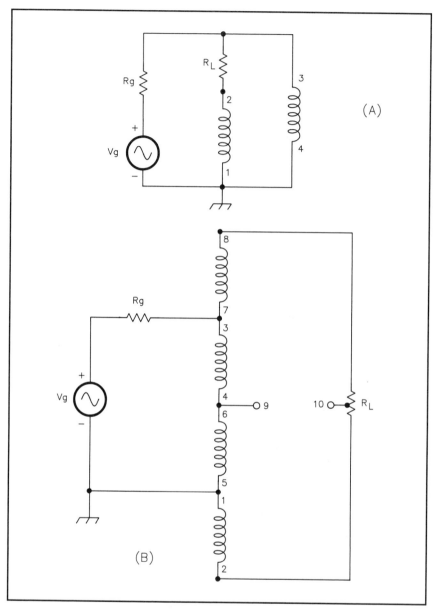

Fig 9-8—Low-frequency models of 1:4 baluns: (A) Ruthroff model and (B) Guanella model.

Ruthroff unun which was described in Chapter 1. On the other hand, the Guanella 1:4 balun (Fig 9-7B) (which appeared in the literature 15 years earlier) obtains a doubling of the input voltage, V_1, by simply adding the outputs of two transmission lines. These coiled transmission lines are connected in parallel at the low-impedance side and in series at the high-impedance side. Three important distinctions are noted in these two approaches, when using a floating load, and they are:

1) The Ruthroff balun only works in one direction. The high-impedance side (on the right in Fig 9-7A) is always the balanced side. The Guanella balun, on the other hand, is bilateral. It can work as well in either direction, depending on which terminal (1, 5 or 2 in Fig 9-7B) is grounded. Therefore, a Guanella balun can easily be designed to match a 50-Ω coaxial cable to a 12.5-Ω balanced load.

2) The high-frequency response of the Ruthroff balun is considerably less than that of the Guanella balun since it adds a delayed voltage to a direct voltage. At the frequency in which the delay is 180 degrees, the output of the Ruthroff balun (and the unun as well) is zero. Since the Guanella balun adds two voltages of equal phases, the upper frequency limit is mainly determined by the parasitics due to the coiling of the transmission lines. If the transmission lines are effectively terminated in their characteristic impedances ($R_L / 2 = Z_0$), and parasitics are minimized, then the impedance ratio of the Guanella balun is essentially frequency independent.

3) The Guanella concept is a modular approach that can be extended to yield higher impedance ratios. Three transmission lines can easily be connected in a parallel-series arrangement resulting in a broadband 1:9 balun, four transmission lines would result in a 1:16 balun, and so on. With a practical limit of about 200 Ω, which is obtainable for the characteristic impedance of a coiled transmission line, efficient, broadband baluns matching 40 Ω to 1000 Ω are possible. The Ruthroff balun cannot possibly compete in the arena.

Another interesting comparison is seen in their low-frequency models shown in Fig 9-8. If the total number of turns in the two coils of the Ruthroff balun is the same as the total in the four coils of the Guanella balun, and if a single toroid is used (making sure the windings in the Guanella balun are in series-aiding, see Chapter 13), then the low-frequency responses are identical. If the two transmission lines of the Guanella balun are wound on separate toroids and the total number of turns on each toroid equals the total number on the Ruthroff balun, then the low-frequency response of the Guanella balun is better by a factor of two. In either case, the single toroid or the two-toroid, the

Guanella balun has a much higher frequency response since it adds in-phase voltages.

(B) R_L Grounded at Mid-Point

A different condition arises when the load, R_L, is center-tapped to ground. The low-frequency response of the Ruthroff balun is essentially unchanged. But the high-frequency response, unexpectedly, takes on the nature of a Guanella balun; that is, measurements show that the high-frequency response is vastly improved, indicating that two in-phase voltages are now being summed. This could be of interest in designing hybrid transformers or 1:4 baluns with loads center-tapped to ground. On the other hand, the Guanella balun is quite seriously affected in its low-frequency response when a single core is used. From Fig 9-8, it can be seen that winding 1-2 has $R_L / 2$ directly across its terminals. Thus, the reactance of winding 1-2, alone, should be much greater than $R_L / 2$. This "loading" of winding 1-2 is also reflected into the other three windings because of the tight magnetic coupling at low frequencies. In this case, a series 1:1 balun is necessary in order to restore the low-frequency response to the "floating" R_L condition. If two separate toroids are used with the Guanella 1:4 balun, then center-tapping R_L to ground has only a small effect.

The examples of 1:4 baluns which now follow are broken down according to impedance levels. Comparisons are shown between Ruthroff and Guanella baluns, as well as between ferrite and powdered-iron core baluns. The latter comparison is particularly directed toward the use of 1:4 baluns in antenna tuners. This comparison does not answer all of the questions related to antenna tuners, since the total solution involves the design of the L-C network and the length and character of the feed line. The complete design of antenna tuners is beyond the scope of this book. The examples presented in this section on low-impedance baluns only use Guanella's approach. Since his baluns are bilateral, excellent 1:4 baluns matching 50-Ω coax to 12.5 Ω (balanced) are readily designed. These baluns should find use in matching coax directly to Yagi beam antennas without delta matches, hairpins and so on. Other baluns having ratios less than 1:4, which can be used for Yagi beams of various element spacings as well as for quad antennas, are described later in the chapter.

Sec 9.3.1   50:200-Ω Baluns

Fig 9-9 shows two 1:4 baluns designed to match into "floating" balanced loads in the range of 150 to 300 Ω. Their best high-frequency responses occur when matching 50-Ω coax into 200 Ω. The characteristic impedances of their windings are about 100 Ω. The parameters for these transformers are as follows:

A) *Left:*

This Ruthroff balun has 15 bifilar turns of no. 14 wire on a 3-inch OD, 4C4 toroid ($\mu = 125$). Each wire is covered with 17-mil wall Teflon tubing. The reactance, due to 15 turns, is sufficient to allow efficient operation down to 1.5 MHz. The coax connector is on the low-impedance (unbalanced) side.

B) *Right:*

Each of the two transmission lines of this Guanella balun has 9 bifilar turns of the same wire and insulation as above. The toroid is also the same as above. Since the total number of turns is now 18, an extra

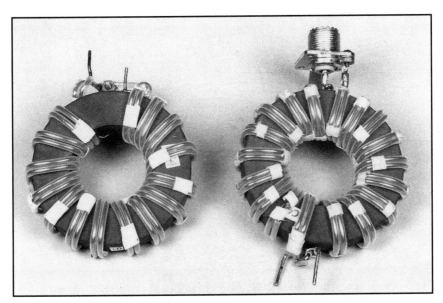

Fig 9-9—Two 1:4 baluns designed to match 50 Ω unbalanced to 200 Ω balanced. On the left is a Ruthroff design; on the right, a Guanella design.

margin of 50% exists in the low-frequency response over the Ruthroff balun. The coax connector is also on the low-impedance side.

Fig 9-10 shows a comparison in the frequency response between these two baluns for three different values of the load, R_L. These results, which highly favor the Guanella balun, were obtained from the simple resistive bridge described in Chapter 12 and really don't tell the whole story. At the optimum impedance level of 50:200-Ω (for both baluns), the Ruthroff balun showed appreciable phase shift beyond 15 MHz, while the Guanella balun showed virtually no phase shift even up to 100 MHz. The phase shift is indicated by the depth of the null when

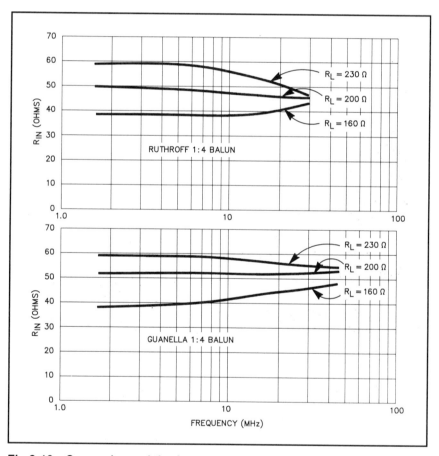

Fig 9-10—Comparison of the frequency response between the two 1:4 baluns shown in Fig 9-9.

compared to that of a carbon resistor (of comparable value) connected directly to the bridge. And finally, the 4C4 toroid from Ferroxcube was selected since it does not have the failure mechanism with high flux density that practically all of the other ferrites possess. It is the ferrite of choice where possible abuse (as in an antenna tuner) can take place.

Fig 9-11 is a photograph of two other 1:4 baluns which supply further information regarding their use in antenna tuners. These two baluns were also designed to operate best at about the 50:200-Ω level. Their parameters are as follows:

A) *Left:*

This Ruthroff balun approximates the 1:4 balun used in some of the antenna tuners. It has 16 bifilar turns of no. 14 wire on two T-200, 2-inch OD, powdered-iron toroids ($\mu = 10$). One of the wires is covered with 17-mil wall Teflon tubing. Since the characteristic impedance of this coiled transmission line is 90 Ω, the impedance level for best high-frequency response is 45:180-Ω.

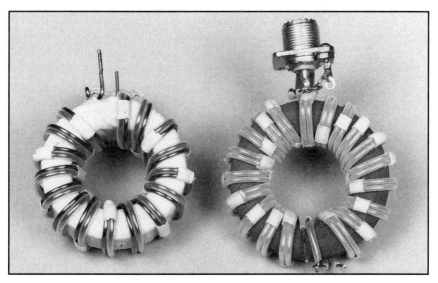

Fig 9-11—Two 1:4 baluns used in the study on baluns for antenna tuners. The balun on the left, which uses two powdered-iron toroids, approximates the one used in some antenna tuners. The balun on the right uses a single ferrite toroid.

B) *Right:*

This Guanella balun has 10 bifilar turns of no. 16 wire in each of the two transmission lines (thus 20 turns when considering the low-frequency response) on a 2.4-inch OD, no. 61 toroid ($\mu = 125$). The wires are covered with 17-mil wall Teflon tubing. The characteristic impedance is 105 Ω and the best high-frequency response occurs at about the 52.5:210-Ω level. This transformer is capable of handling 1 kW of continuous power.

Fig 9-12 shows the frequency responses of these two baluns with various (floating) terminations; it not only demonstrates the superiority of the Guanella balun with a ferrite core, but also the danger of using a

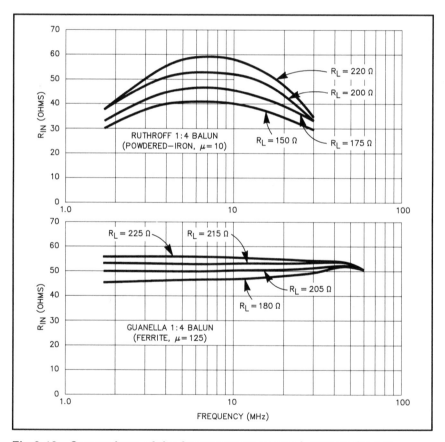

Fig 9-12—Comparison of the frequency response between the 1:4 baluns shown in Fig 9-11.

powdered-iron balun in an antenna tuner. It also shows that the powdered-iron balun starts falling off, at the low end, at around 7 MHz. This means there is insufficient reactance in the windings to prevent a sizable shunting effect. Below 7 MHz, the transformer becomes inductive. This condition allows for flux in the core. Further, this inductance can become part of the tuned L-C network, resulting in very high currents and flux densities. The Guanella (ferrite) balun in Fig 9-12 shows an exceptional response from 1.7 MHz to 60 MHz. This balun, which uses the popular 2.4-inch OD, no. 61 toroid ($\mu = 125$), should be investigated for possible use in antenna tuners.

Sec 9.3.2   75:300-Ω Baluns

The 75:300-Ω balun has been quite popular because of its ability to match 75-Ω coax to the resonant impedance of folded dipoles (which is about 300 Ω). With the use of a very broadband 1:1.56 unun (see Chapter 7) in series, 50-Ω coax can easily be matched to 300 Ω. Fig 9-13 shows three broadband Guanella 1:4 baluns designed to match into balanced, floating loads. The balun in the center is specifically designed to match 75 Ω to 300 Ω. The balun on the left, although optimized for the 50:200-Ω level, is still able to cover three of the Amateur Radio bands at the 75:300-Ω level because of the short lengths of its transmission lines. The balun on the right is a 25:100-Ω design which will be described in the next subsection. The parameters for the two Guanella baluns on the left in Fig 9-13 are:

A) *Left:*

This Guanella balun, which is optimized at the 50:200-Ω level, has 7 bifilar turns on the two transmission lines. The wire is no. 16 and is covered with 17-mil wall Teflon tubing. The core is a 1¾-inch OD, 250L toroid ($\mu = 250$). The response at the 75:300-Ω level is flat from 1.5 MHz to 10 MHz. Beyond 10 MHz, the impedance ratio increases and becomes complex. This balun performs much better at this level than the larger Guanella balun on the right in Fig 9-9 because of its much shorter transmission lines. At the 50:200-Ω level, this balun is essentially flat from 1.5 MHz to 100 MHz. Also, at the 25:100-Ω level, this balun still covers the 1.5-MHz to 10-MHz range. The power rating is 1 kW of continuous power at the three impedance levels.

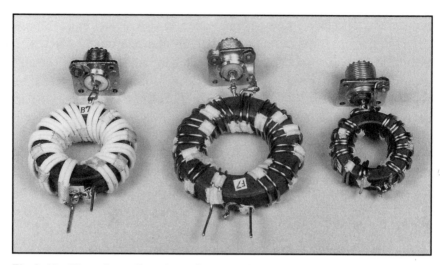

Fig 9-13—Three broadband 1:4 Guanella baluns. The one on the left is designed to match 50 Ω to 200 Ω. The center one is designed to match 75 Ω to 300 Ω. The one on the right is designed to match 25 Ω to 100 Ω.

B) *Center:*

This Guanella balun is optimized for the 75:300-Ω level. The spacing between the no. 16 wire (ladder line) is approximately ⅛-inch. The spacers are sections of Scotch no. 27 glass tape. Each transmission line has 7 bifilar turns on the 2.4-inch OD, no. 64 toroid ($\mu = 250$). The response is flat from 3.5 MHz to 50 MHz. If operation down to 1.5 MHz is required, then a 2.62-inch OD, K5 toroid ($\mu = 290$) and 9 bifilar turns is recommended. A permeability of 125 would probably raise the efficiency from 97% to 98%. A permeability of 40 would raise it to 99%. The same comment can be made for (A) above. But lowering the permeability affects (negatively) the low-frequency response proportionally.

Sec 9.3.3   25:100-Ω Baluns

Although not nearly as popular as the baluns described previously, the 25:100-Ω balun does have some applications. For example, if it is in series with a 50:25-Ω unun (see Chapter 7), then a broadband match can be made from a 50-Ω coax to the balanced and floating impedance of a quad antenna. This compound arrangement can also be done on a single toroid and is described in Sec 9.5.2. The balun on the right in Fig 9-13, which is actually optimized at the 27.5:110-Ω level, has an essentially flat 1:4 ratio from 1.5 MHz to 100 MHz with loads varying from 90 Ω to 120 Ω. It has 6 bifilar turns of no. 16 wire (held closely together) in each of its two transmission lines. The core is a 1½-inch OD, no. 64 toroid (μ = 250). The first version of this balun had the same number of bifilar turns, but they were spaced much closer to each other. The best high-frequency response occurred at the 25:100-Ω level. This balun showed appreciable phase shift with a resistive bridge (and hence roll-off) at 50 MHz. By simply increasing the spacing between adjacent bifilar turns (particularly on the inside diameter of the toroid), the high-frequency response more than doubled. Virtually no phase shift was observed at 100 MHz, which was the limit of the bridge (see Chapter 12). This was a clear demonstration of the very-high-frequency capability of the Guanella balun when the parasitics are considerably reduced. The Ruthroff balun, because it adds a delayed voltage to a direct voltage, cannot approach this performance when the load is "floating."

Sec 9.3.4   12.5:50-Ω Baluns

Very broadband baluns matching 50-Ω coaxial cable to balanced loads of 12.5 Ω (floating or grounded at their mid-points) are readily made using Guanella's approach. In this case, the high side of the transformer is grounded; that is, terminal 2 in the Guanella balun of Fig 9-7B instead of terminal 1. These transformers can be designed to maintain their 1:4 impedance ratios over a very wide bandwidth with loads that vary from 9 Ω to 15 Ω. This is the range for many Yagi beam antennas. These baluns can also be made to handle more than a kW of continuous power. By using low-impedance coaxial cable (Sec 4.3) or polyimide-coated wire like ML or H Imideze, they can withstand several thousand volts without breakdown. Three of these baluns are shown in Fig 9-14. The coax connectors are now on the high side of the baluns. The parameters for these three baluns are as follows:

A) *Left:*

12 bifilar turns of no. 14 wire, tightly wound on ⅜-inch diameter rods, 4 inches long, of μ = 125 ferrite. The impedance ratio, when matching 50 Ω (unbalanced) to 12.5 Ω (balanced and floating), is essentially flat from 1.5 MHz to over 30 MHz. With the center of the 12.5-Ω load grounded, the range is 3 MHz to over 30 MHz. The power

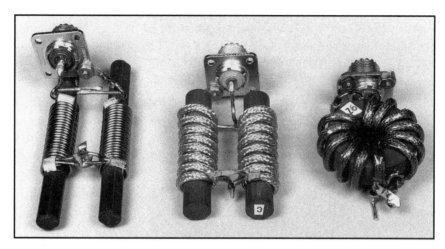

Fig 9-14—Three broadband 1:4 Guanella baluns designed to match 50-Ω coaxial cable to balanced loads of 12.5 Ω.

rating is 1 kW of continuous power. If ½-inch diameter rods were used, then 10 bifilar turns would give the same performance. The length of the rods, which is not especially critical, can vary from 3 to 4 inches.

B) *Center:*

7½ turns of low-impedance coaxial cable on ½-inch diameter, 2½-inch long rods of μ = 125 ferrite. The inner conductor of no. 12 wire has 2 layers of Scotch no. 92 tape. The outer braid, which is unwrapped, is from RG-122/U cable (or equivalent). At the 11.75:47-Ω level (which is the optimum level), the response with a floating load is flat from 3.5 MHz to well over 30 MHz. With the center of the load grounded, it is flat from 7 MHz to well over 30 MHz. These numbers are pretty well duplicated at the 12.5:50-Ω level. The power rating is over 2 kW of continuous power. The voltage breakdown is in excess of 3000 V if ML or H Imedeze wire is used.

C) *Right:*

Two coaxial cable transmission lines of 6 turns each on a single 1½-inch OD, 4C4 toroid (μ = 125). The inner conductor is no. 14 wire with 6 layers of Scotch no. 92 tape. The outer braid, which is from RG-122/U cable (or equivalent), is tightly wrapped with Scotch no. 92 tape. When matching 50 Ω (unbalanced) to 12.5 Ω (balanced and floating), the response is flat from 1.5 MHz to 50 MHz. At the 11.5:36-Ω level (which is the optimum impedance level), it is flat from 1.5 MHz to well over 65 MHz. By using permeabilities of 250 to 300, and fewer turns (like 3 or 4), this balun could also cover an appreciable portion of the VHF band. To repeat, since a single toroid is used, this balun is not designed to match into a load which is center-tapped to ground.

Sec 9.4 The 1:9 Balun

When matching a 50-Ω coax down to a balanced load of 5.56 Ω or up to a balanced load of 450 Ω, the Guanella balun is the transformer of choice. There is little doubt that his baluns offer the widest bandwidths under these two very different conditions. Further, his modular concept (that is, adding transmission lines in parallel-series arrangements) offers the highest efficiency at high impedance levels. Experiments have shown (see Chapter 11) that efficiency, at least with the Ruthroff unun, decreases as the impedance level increases. With Guanella's approach, each transmission line shares a portion of the load; therefore, his transmission lines can work at lower impedance levels. Also, the longitudinal gradients are less with his transformers.

Figs 9-15 and 9-16 are photographs of three 1:9 baluns designed for these two extremes in impedance levels. The balun in Fig 9-15 is designed to match 50 Ω to 5.56 Ω. This transformer could be used to match 50-Ω coaxial cable directly to short-boom, four-element Yagi

Fig 9-15—A low-impedance 1:9 Guanella balun designed to match 50-Ω coaxial cable to a balanced load of 5.56 Ω.

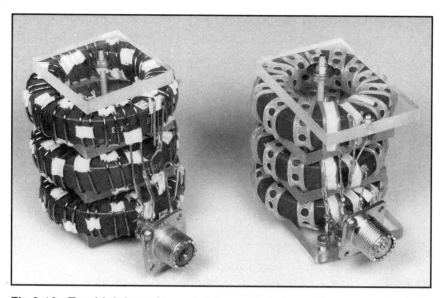

Fig 9-16—Two high-impedance 1:9 Guanella baluns. The one on the left is designed to match 50 Ω to 450 Ω. The one on the right is designed to match 66.7 Ω to 600 Ω.

beams with resonant impedances of about 6 Ω. Fig 9-16 is a photograph of two 1:9 baluns capable of broadband operation from 450 Ω to 600 Ω. By using four transmission lines, instead of three, baluns capable of matching 50 Ω to 3.125 Ω or to 800 Ω become readily available. The parameters for the three baluns in Figs 9-15 and 9-16 are as follows:

A) *Fig 9-15:*

This low-impedance 1:9 balun has 9½ turns of low-impedance coax (Z_0 = 13 Ω) on each of the three ferrite rods (μ = 125). The rods have a diameter of ½-inch and a length of 4 inches. The schematic is shown in Fig 9-17A. The inner conductors are no. 12 wire with two layers of Scotch no. 92 tape. The outer braids (unwrapped) are from RG-122/U cable (or equivalent). At the 5.56:50-Ω level, the impedance ratio (with the load floating) is constant from 1.5 MHz to over 30 MHz. The power rating is in excess of 2 kW of continuous power. With ML or H Imideze wire, the voltage breakdown is in excess of 3000 V. Although more awkward to construct, 4 turns of the same coax on toroids with permeabilities of 250 to 300 would yield a 1:9 balun with much greater bandwidth. And lastly, a broadband 1:16 balun could be constructed with

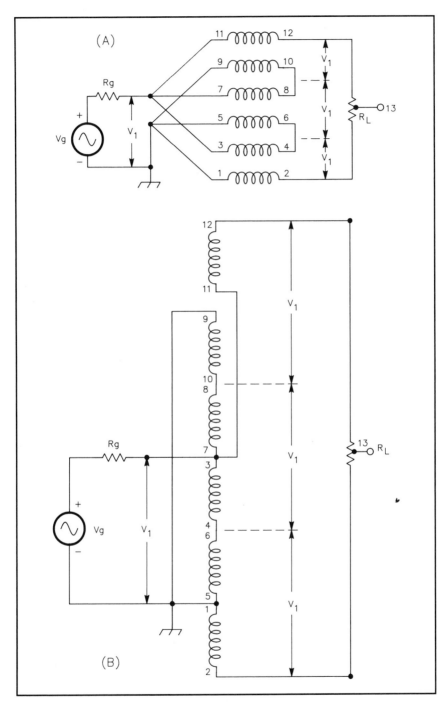

Fig 9-17—Models of Guanella's 1:9 balun: (A) high-frequency model and (B) low-frequency model. It is assumed that $Z_0 = R_L / 3$ and therefore V_2, the output of each transmission line, equals V_1.

four coaxial cables using no. 10 wire with one layer of Scotch no. 92 tape for the inner conductor. The characteristic impedance of this coax would be about 9 Ω. This balun would match 3.125 Ω to 50 Ω.

B) *Fig 9-16, left:*

This high-impedance 1:9 step-up balun has 16 bifilar turns of no. 18 wire (spaced about 3⁄16 inch) on each of the three 2.62-inch OD, K5 toroids (μ = 290). In the schematic of Fig 9-17, the ground is connected to the low-impedance side (on the left). The characteristic impedance of the transmission lines is 170 Ω and the best high-frequency response occurs at the 61:550-Ω impedance level. At this level, the response is flat from 1.5 MHz to 45 MHz. If no. 16 wire were used, the characteristic impedance would be 150 Ω and the best high-frequency response would occur at the 50:450-Ω level. If a slightly higher efficiency (greater than 97%) is required, then a core permeability of 40 or 125 is recommended. The low-frequency response would be raised (and hence poorer) by a factor of 7 and 2, respectively. The efficiencies would also be raised to 99% and 98%, respectively. It is quite surprising that large structures like these have such wide bandwidths. Adding in-phase voltages (from flat lines) is the key. The power rating is 1 kW of continuous power with either no. 16 or 18 wire. Further, by adding 1:1 baluns (on the low-impedance side) in series, these compound transformers become excellent ununs.

C) *Fig 9-16, right:*

This is another high-impedance 1:9 Guanella balun (even higher than [B] above). Each of the three 2.4-inch OD, no. 64 toroids (μ = 250) has 12 bifilar turns of 300-Ω TV ribbon. Holes are punched in the ribbon to make it easier to tightly wind it around the toroids. Due to the close proximity of the turns on the inside diameters of the cores, the characteristic impedance is lowered to 205 Ω. At the 68.33:615-Ω level, which is optimum, the response is flat from 5 MHz to 40 MHz. By using a larger core, like the 2.62-inch OD, K5 toroid (μ = 290), one to two more turns would be possible, thereby extending the frequency range to 3.5 MHz to 40 MHz. At the 66.67:600-Ω impedance level, the high-frequency response of these transformers is still above 30 MHz. The power rating is at least 500 W of continuous power. Further, by adding a 1:1.36 unun and a 1:1 balun in series (and this can be done with one core, see

Chapters 7 and 8), this compound arrangement becomes an excellent unun for matching 50-Ω (unbalanced) to 600 Ω (unbalanced).

Sec 9.5 Baluns for Yagi, Quad and Rhombic Antennas

To date, the most popular balun for antenna use has been the 1:1 (50:50-Ω, nominally) trifilar design by Ruthroff. It has been used successfully in matching 50-Ω coax to Yagi beams after shunt-fed methods were employed in order to raise the input impedance. It has also found success in matching 50-Ω coax directly to ½-λ dipoles at heights of 0.15 to 0.2 λ, where the resonant impedances are 50 to 70 Ω, respectively (the resonant impedance reaches a peak of about 98 Ω at a height of 0.34 λ). Outside of these two cases, baluns have found very little use in matching 50-Ω coaxial cable to resonant impedances far removed from the "nominal" 50 Ω. For the experimentalists, baluns for the following antennas are offered.

Sec 9.5.1 Yagi Beams

Sec 9.3 and Sec 9.4 described Guanella baluns with ratios of 1:4 and 1:9 which can match 50-Ω coax directly to Yagi beams with balanced and floating impedances of about 9 to 15 Ω and 5 to 8 Ω, respectively. Fig 9-18 shows a photograph of two other baluns capable of matching 50-Ω coax directly to higher-impedance Yagi antennas. On the left in

Fig 9-18—Two fractional-ratio baluns designed to match 50-Ω coaxial cable directly to the inputs of wide-spaced Yagi beams. The balun on the left is designed to match 50 Ω to 20 Ω. The balun on the right is designed to match 50 Ω to 30 Ω.

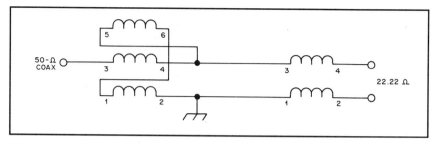

Fig 9-19—Schematic of the fractional-ratio balun designed to match 50-Ω coaxial cable to a balanced and floating load of about 20 Ω. The 1:1 balun on the right uses 22-Ω coaxial cable.

Fig 9-18 is a balun designed to match 50-Ω coax to a balanced (and floating) impedance of about 20 Ω. Its useful impedance range is probably from 16 Ω to 25 Ω. The schematic is shown in Fig 9-19. It is a compound transformer consisting of a step-down (50:22.22-Ω) Ruthroff-type unun in series with a low-impedance coaxial-cable 1:1 (22:22-Ω) Guanella balun. The common core is a 2-inch OD, no. 61 toroid (μ = 125). The unun has 5 trifilar turns of no. 14 wire. The 1:1 coaxial-cable balun also has 5 turns. The coaxial cable uses no. 12 wire with two layers of Scotch no. 92 tape for the inner conductor. The outer braid, which is left untaped, is from RG-122/U. At the 50:22.22-Ω level, the response is flat from 3.5 MHz to well beyond 30 MHz. The power rating is in excess of 1 kW of continuous power. If 160-meter operation is desired, then a core with a permeability of 250 to 300 is recommended.

On the right in Fig 9-18 is a balun designed to match 50-Ω coax to a balanced and floating impedance of about 30 Ω. Its useful imped-

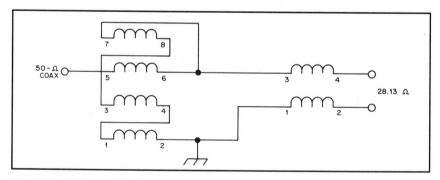

Fig 9-20—Schematic of the fractional-ratio balun designed to match 50-Ω coaxial cable to a balanced and floating load of about 30 Ω. The 1:1 balun on the right uses 30-Ω coaxial cable.

ance range is probably from 25 Ω to 35 Ω. It is also a compound transformer using a step-down (50:28.13-Ω) Ruthroff-type unun in series with a low-impedance coaxial-cable 1:1 (30:30-Ω) Guanella balun. Fig 9-20 shows the schematic. The common core is a 2.4-inch OD, no. 64 toroid (μ = 250). The unun, which has an impedance ratio of 1.78:1, has 6 quadrifilar turns. Winding 5-6, in Fig 9-20, is no. 14 wire, and the other three are no. 16 wire. The 1:1 coaxial-cable balun also has 6 turns. The inner conductor of the coaxial cable is no. 14 wire with two layers of Scotch no. 92 tape, followed with two layers of Scotch no. 27 glass tape. The outer braid (untaped) is from RG-122/U cable. At the 50:28.13-Ω level, the response is flat from 1.5 MHz to 50 MHz. The power rating is in excess of 1 kW of continuous power. This transformer, as well as the one above, could also have been constructed with two separate cores.

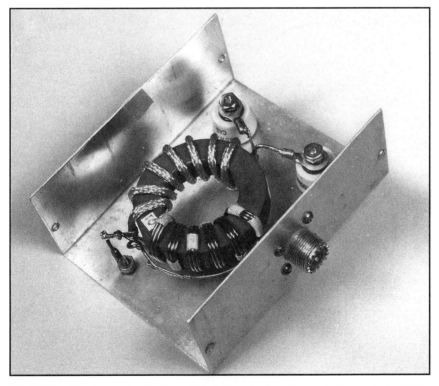

Fig 9-21—Photograph of a low-impedance, step-down balun mounted in a Minibox. The balun uses the schematic of Fig 9-19 and matches 50-Ω coaxial cable to a balanced and floating load of about 20 Ω.

Fig 9-21 shows a balun, mounted in a Minibox, which is similar to the one on the left in Fig 9-18. It is also designed to match 50-Ω coax to about a 20-Ω balanced and floating load. It also uses the schematic of Fig 9-19. This balun, which uses a 2.4-inch OD, Q1 toroid ($\mu = 125$), has 6 trifilar turns of no. 14 wire for the unun and 7 turns of 22-Ω coaxial cable for the 1:1 Guanella balun. At the 50:22.22-Ω level, the response (Sec 4.3) is flat from 1.5 MHz to well over 30 MHz. Its power rating is also 1 kW of continuous power.

Sec 9.5.2 Quad Antennas

The quad antenna generally has a balanced (and floating) resonant impedance in the range of 100 to 120 Ω. This antenna also lends itself readily to a compound balun. Several approaches can be used; for example, a 1:2 step-up unun (50:100-Ω) followed by a 1:1 balun (100:100-Ω), or a 2:1 step-down unun (50:25-Ω) followed by a 1:4 step-up balun (25:100-Ω). The latter approach was tried by the author and is reported in this subsection. Fig 9-22 shows a photograph of a

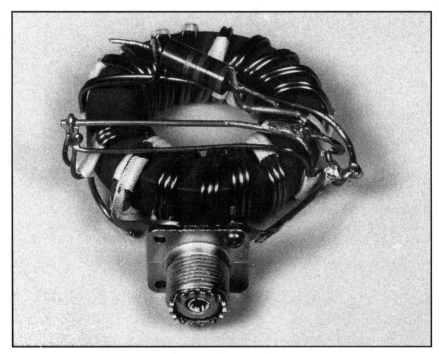

Fig 9-22—A 1:2 step-up balun designed to match 50-Ω coaxial cable to the balanced input impedance of a quad antenna.

compound balun using a single 2.4-inch OD, no. 61 toroid ($\mu = 125$). The schematic is shown in Fig 9-23. It uses a tapped-trifilar step-down unun in series with a Ruthroff 1:4 balun. The Guanella 1:4 balun, although possessing a better high-frequency response, was not used since it did not lend itself as readily to a single core. If a much wider bandwidth is required, then two separate cores, with the 1:4 balun using Guanella's approach, is recommended. The unun in Fig 9-23 has 6 trifilar turns. Winding 3-4 is no. 14 wire and is tapped at one turn from terminal 3. The other two windings are no. 16 wire. With the input connection to the tap, the impedance ratio is 2:1. The Ruthroff 1:4 balun uses 10 bifilar turns of no. 14 wire. This compound balun, 50-Ω coax to 100-Ω (balanced), is flat from 3.5 MHz to 30 MHz. The response is quite the same in matching 60 Ω (unbalanced) to 120 Ω (balanced). With the input connection directly to terminal 3, a similar response is obtained at the 50:90-Ω impedance level. If 160-meter operation is also desired, then a toroid with a permeability of 250 to 290 is recommended. The power rating is 1 kW of continuous power.

Sec 9.5.3   Rhombic Antennas

Compound baluns also lend themselves to matching 50-Ω coax to the balanced (and floating) resonant impedances of V and rhombic antennas. These impedances are generally on the range of 500 to 700 Ω. Sec 9.4 described some compound baluns using ununs in series with Guanella baluns, yielding wideband responses over this impedance range. This subsection presents an earlier approach by the author for a

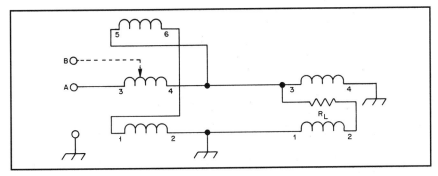

Fig 9-23—Schematic of the 1:2 step-up balun shown in Fig 9-22. With the input to terminal B, the impedance ratio is 1:2 (50:100-Ω). With the input to terminal A, the impedance ratio is 1:1.78 (50:90-Ω).

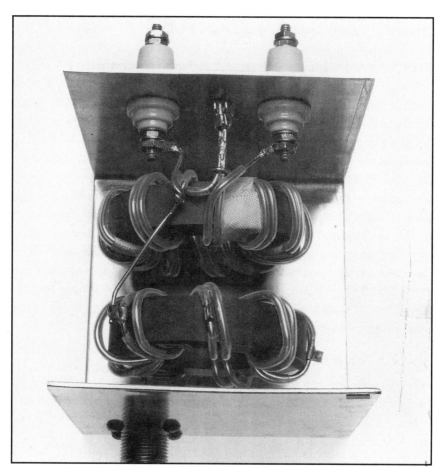

Fig 9-24—A 1:12 balun designed to match 50-Ω coaxial cable to a balanced load of 600 Ω.

1:12 balun using only two toroidal transformers. One is a tapped-bifilar Ruthroff unun with a ratio of 1:3 and the other is a Ruthroff 1:4 balun. Fig 9-24 is a photograph of the compound balun. Fig 9-25 shows the performance when matching to a balanced (and floating) load of 600 Ω. In this case, the input impedance is measured as a function of frequency. As shown, a constant ratio is obtained in the frequency range of 7 MHz to 30 MHz. Fig 9-26 shows the schematic for the two series transformers. The transformer on the left has 8 bifilar turns of no. 14 wire on a 2.4-inch OD, Q1 toroid (μ = 125). The wire is covered with 17-mil wall Teflon tubing. The top winding in Fig 9-26 is tapped 6 turns from

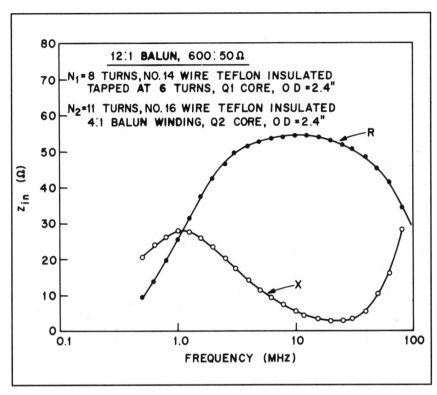

Fig 9-25—Performance of the 1:12 balun of Fig 9-24.

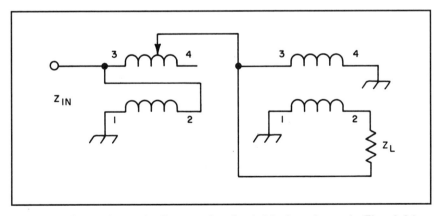

Fig 9-26—The schematic diagram for the 1:2 balun shown in Figs 9-24 and 9-25.

terminal 3, giving a 1:3 step-up ratio. The transformer on the right has 11 bifilar turns of no. 16 wire on a 2.4-inch OD, Q2 toroid ($\mu = 40$). The wire is also covered with 17-mil wall Teflon tubing. This is a 1:4 Ruthroff balun. Although Q2 material has a lower permeability than Q1 material, it was chosen because it has a lower core loss at these high impedance levels (see Chapter 11).

This combination of transformers allows for broadband operation at the high impedance levels of 500 to 700 Ω because of the canceling effects they have in a series configuration. Since the characteristic impedance, Z_0, of the 1:4 balun on the right is only about 130 Ω (it should be 300 Ω for optimum response), the input impedance as seen at its terminals 1-3 is capacitive and the real part is less than $R_L / 4$ (see Figs 1-5 and 1-6). Since the characteristic impedance, Z_0, of the tapped transformer on the left is about 115 Ω, and is greater than would be normally used to match 50 Ω to 150 Ω, it has the opposite effect on its load. It causes the load to look inductive; the reactance of the right-hand transformer is effectively canceled over a large portion of the band. The resistive component is not altered when the characteristic impedance is greater than would normally be used.

Even though the compound baluns described in Sec 9.4 (using Guanella's approach) have the potential for much wider bandwidths, the two octaves achievable by the simple schematic of Fig 9-26 should prove to be quite useful.

Chapter 10

Multimatch Transformers

Sec 10.1 Introduction

Multiband vertical antennas have enjoyed considerable popularity with radio amateurs (including the author) because of their low angle of radiation and flexibility in changing bands without any form of external switching. Most of these multiband verticals have high degrees of inductive loading, and thus very narrow bandwidths on the 80- and 160-meter bands.

In order to obtain greater bandwidths on these bands and still maintain a multiband antenna system with a high degree of vertical radiation, a four-band (10-40 meters) trap vertical was connected, by the author, in parallel with sloper and inverted-L antennas for the 80- and 160-meter bands. This was done over a low-loss ground system of 100 radials, each about 50 to 60 feet in length. Impedance measurements (with the simple resistive bridge described in Chapter 12) showed that the sloper and inverted-L antennas had resonant impedances of 12 to 25 Ω, depending upon: (A) the angle of slope (of the sloper), (B) the height of the vertical portion of the inverted L, and (C) the interaction between the 80- and 160-meter antennas. Since the inverted-L antennas were mounted 8 to 12 inches from the trap vertical, little difference was noted in the input impedance of the trap vertical (and hence performance). Slopers had very little effect since the capacitive coupling was minimal. Instead of using separate feed lines and matching transformers, or a single feed line and relays which switch to the appropriate matching transformers, parallel transformers were investigated for possible use. Two step-down transformers (with different ratios) were connected in parallel on their 50-Ω sides and to various antennas on their output sides. Since the mutual coupling of parallel-connected transmission line transformers was found to be minimal (like putting a short length of unterminated transmission line across their inputs), this arrangement pretty much duplicated the well-known technique of feeding parallel-connected dipoles (for different bands) with a single coax. This process worked well for two different impedances, but left something to be desired if three or four broadband ratios were required.

This then led to an investigation for obtaining two broadband ratios from a single transformer. Transformers capable of supplying either 1:1.5 and 1:3 or 1:2 and 1:4 ratios were then successfully designed. Again, by connecting these two transformers at their 50-Ω sides, which were the high-impedance sides, four broadband ratios now became available. Although this technique was only used for matching 50-Ω coax to lower impedances, it should also work in matching 50-Ω coax to a variety of impedances, high and low.

This chapter describes the various broadband, multimatch (unbalanced to unbalanced) transformers used with the author's ground-fed multiband antenna systems, as well as the general multimatch transformer described in the first edition of this book.

Sec 10.2   Dual-Output Transformers

Chapters 7 and 8 introduced the concept of higher-order windings and how broadband ratios of less than 1:4 could be obtained with trifilar, quadrifilar, and so on, transformers. Of particular significance was the transposition of the various windings in order to obtain optimum characteristic impedances for either rod or toroidal transformers. Most of these transformers could yield more than one impedance ratio by either tapping a winding or by direct connections to the terminals of the inner windings. Generally these transformers were optimized for a single impedance ratio. This section introduces the concept of transposing the windings such that two broadband ratios become available. As will be seen, the schematics are considerably different from those of the two earlier chapters. In general, these transformers do not quite exhibit the high-frequency response, for both ratios, of a single-ratio transformer. This is because it is difficult to optimize equally the characteristic impedances of the windings for two different broadband ratios. But in most cases, the two impedance ratios can be found to be constant from 1.5 MHz to 30 MHz. In many cases, the high-frequency response of one of the ratios easily exceeds 30 MHz.

Sec 10.2.1 1:1.5 and 1.3 Ratios

Although the actual ratios obtained by the following transformers are 1:1.56 and 1:2.78, respectively, very little difference will be noted from ratios of 1:1.5 and 1:3 when matching to antennas. This is because of the variation of their input impedances with frequency. A slight shift in the best match-point (lowest VSWR) with frequency might be observed. Fig 10-1A shows the basic schematic of a dual-output (on the left side), quintufilar, unbalanced-to-unbalanced transformer. Terminal C is the high-impedance side. Figs 10-1B and 10-1C, which are toroidal and rod versions, are specifically designed to match 50-Ω coax to lower impedances. The two broadband impedance ratios, which are the same for all three transformers, can best be determined from the basic schematic of Fig 10-1A. They are as follows:

A) *1:1.56 ratio:*

If an input voltage V_1 is connected from terminal A to ground, the four transmission lines have $V_1 / 4$ on their inputs. The terminals 2, 4, 6 and 8 are all "boot-strapped" by a potential of $V_1 / 4$ by their connections to the odd-numbered terminals on the left. The voltage between terminals 10 and 8 came from $V_1 / 4$, on the left, which traversed a transmission line. Thus, the voltage at terminal C becomes $5/4V_1$ and the impedance ratio becomes $(V_{out} / V_1)^2 = (5 / 4)^2 = 1.56$. The top

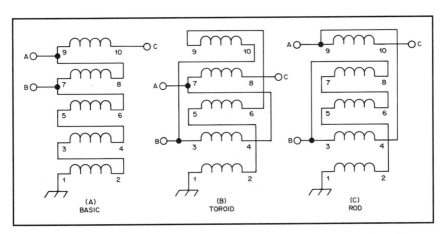

Fig 10-1—Schematic diagrams of the basic version of the dual- ratio quintufilar transformer and the transposed-winding versions favoring the toroid or the rod when matching 50-Ω coax at terminal C to 32 Ω at terminal A or 18 Ω at terminal B.

winding, 9-10, carries four-fifths of the current into terminal A, and the other four only one-fifth of the current.

B) *1:2.78 ratio:*

If an input voltage V_1 is connected from terminal B to ground, then the three bottom transmission lines have $V_1 / 3$ on their inputs. Terminals 2, 4 and 6 are now "boot-strapped" by a potential of $V_1 / 3$ by their connections to the odd-numbered terminals on the left side. The voltage between terminals 8 and 6 came from $V_1 / 3$, on the left, which traversed a transmission line. The voltage between terminals 10 and 8 came from the same voltage which traversed the transmission line twice. Thus the output at C becomes $5/3V_1$. The impedance ratio then becomes $(V_{out} / V_1)^2 = (5 / 3)^2 = 2.78$. The two top windings in Fig 10-1A carry three-fifths of the current into terminal A, and the other three only two-fifths.

Fig 10-2 is a photograph of three low-impedance, toroidal versions (Fig 10-1B) of the quintufilar, dual-output transformer. The chassis connectors are all on the high-impedance sides of the transformers. The parameters of these three transformers are as follows:

A) *Left:*

5 quintufilar turns on a 1½-inch OD, no. 64 toroid ($\mu = 250$). Windings 7-8 and 3-4 are no. 14 wire and the other three are no. 16 wire.

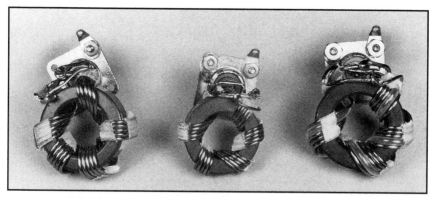

Fig 10-2—Three low-impedance, toroidal versions (Fig 10-1B) of the quintufilar, dual-output unun transformer.

When the input is connected to terminal A (the 1:1.56 ratio), the response is flat from 1.5 MHz to 30 MHz at the 32:50-Ω level. At the optimum impedance level of 29:45-Ω, the response is flat from 1.5 MHz to 45 MHz. When the input is connected to terminal B (the 1:2.78 ratio), the response is flat from 1.5 MHz to 45 MHz at the 18:50-Ω level. At the optimum impedance level of 20:56-Ω, the response is flat from 1.5 MHz to over 50 MHz. The power rating with either ratio is 1 kW of continuous power.

B) *Center:*

5 quintufilar turns on a 1¼-inch OD, K5 toroid (μ = 290). Windings 7-8 and 3-4 are no. 16 wire and the other three are no. 18 wire. When the input is connected to terminal B (the 1:2.78 ratio), the response is flat from 1 MHz to 45 MHz at the 18:50-Ω level. This is at the optimum impedance level. When matching at the 32:50-Ω level, using terminal A, the response is flat from 1 MHz to well over 50 MHz. This is also at the optimum impedance level. This transformer would easily handle the power from any popular transceiver. Its power rating for either ratio is in excess of 200 W of continuous power.

C) *Right:*

This dual-output transformer was designed for 75-Ω operation (on the high side). It has 4 quintufilar turns on a 1½-inch OD, no. 64 toroid (μ = 250). Windings 7-8 and 3-4 are no. 14 wire, and the other three are no. 16 wire. Winding 7-8 is covered with 17-mil wall Teflon tubing. When using terminal A (the 1:1.56 ratio) and matching 50 Ω to 75 Ω, the response is flat from 1.5 MHz to 30 MHz. It is also the same at the 32:50-Ω level. The optimum response occurs at the 38.5:60-Ω level. Here, it is flat from 1.5 MHz to over 50 MHz. When using terminal B (the 1:2.78 ratio) and matching 27 Ω to 75 Ω, the response is flat from 1.5 MHz to well over 50 MHz. This ratio of 1:2.78 also works well at the 18:50-Ω level. At this level the ratio is constant from 1.5 MHz to 30 MHz. The reason for the good performance over these wide impedance levels is because of the short lengths of the transmission lines—only 7½ inches long. The power rating with either ratio is 1 kW of continuous power.

Fig 10-3 is a photograph of two low-impedance, rod versions (Fig 10-1C) of dual-output ununs. A third input connection is also made

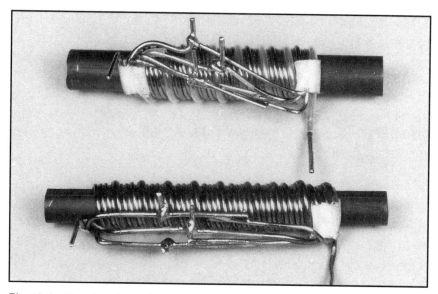

Fig 10-3—Two low-impedance, rod versions (Fig 10-1C) of the quintufilar, dual-output unun transformer.

to terminal 7, yielding a third ratio (but with very much less bandwidth) of 1:6.25. Both of these transformers are rated at 1 kW of continuous power with all ratios. The parameters for these two transformers are as follows:

A) *Top:*

7 quintufilar turns on a ½-inch diameter, 3.75-inch long ferrite rod (μ = 125). Windings 9-10 and 3-4 (in Fig 10-1C) are no. 14 wire. The other three are no. 16 wire. Winding 9-10 is covered with 17-mil wall Teflon tubing. When using terminal A (the 1:1.56 ratio) and matching 32 Ω to 50 Ω, the response is flat from 1.5 MHz to about 30 MHz. At 30 MHz, the ratio becomes a little less than 1:1.56. This 32:50-Ω level is the optimum impedance level. When matching at the 18:50-Ω level, using Terminal B in Fig 10-1C, the response is flat from 1.5 MHz to 30 MHz. At the optimum impedance level of 21.5:60-Ω, it is flat from 1.5 MHz to over 45 MHz. When matching with the 1:6.25 ratio (connecting the input to terminal 7) at the 8:50-Ω level, the response is flat from 1.5 MHz to 10 MHz.

B) *Bottom:*

9 quintufilar turns on a ⅜-inch diameter, 3.75-inch long ferrite rod (μ = 125). Windings 9-10 and 3-4 (in Fig 10-1C) are no. 14 wire. The other three are no. 16 wire. Winding 9-10 is covered with four layers of Scotch no. 27 glass tape (14 mils of insulation). This transformer also has a third input at terminal 7. The response and power rating pretty much duplicate that of the ½-inch diameter rod transformer in (A) above.

Fig 10-4 is a photograph of two dual-output, broadband transformers enclosed in Miniboxes (with appropriate brackets and feedthrough insulators) for mounting at the base of verticals and/or slopers and inverted-L antennas. The rod transformer is identical in construction, performance and power rating to the top transformer in Fig 10-3. The only difference is that winding 9-10 (in Fig 10-1C) is covered with two layers of Scotch no. 27 glass tape (14 mils of insulation) instead of 17- mil wall Teflon tubing. The performance of the toroidal transformer (on the left in Fig 10-4) is identical to the toroidal transformer on the left in Fig 10-2. The power rating is also the same. The only difference in construction is that the core in Fig 10-4 is a 1½-inch OD, 4C4 toroid

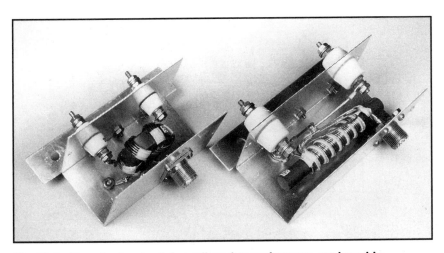

Fig 10-4—Two dual-output, broadband transformers enclosed in Miniboxes. The rod transformer is identical in construction and performance to the top transformer in Fig 10-3. The toroidal transformer is identical to the toroidal transformer on the left in Fig 10-2.

(μ = 125). Since its cross-sectional area is about twice that of the no. 64 toroid (μ = 250) used in Fig 10-2, the low-frequency responses are practically identical. The same is true with the high-frequency responses, since the transmission lines of the two transformers are not significantly different in length.

Sec 10.2.2   1:2 and 1:4 Ratios

In working with ground-fed, multiband antennas systems (over a good ground system) employing combinations of verticals, slopers and inverted-L antennas, a need arose for low-impedance ratios other than 1:1.5 and 1:3. Some antennas had resonant impedances near 12 Ω, and others near 25 Ω. Therefore, a study was undertaken to achieve two broadband ratios of 1:2 and 1:4 with a single core. The quadrifilar transformer was found to yield two broadband ratios of 1:1.78 and 1:4. The 1:1.78 ratio generally satisfies the 1:2 ratio requirement. Fig 10-5 shows the schematic of the basic quadrifilar winding (used for analysis purposes) and the final design (Fig 10-5B) for both rod and toroidal versions. The impedance ratios are determined from Fig 10-5A as follows:

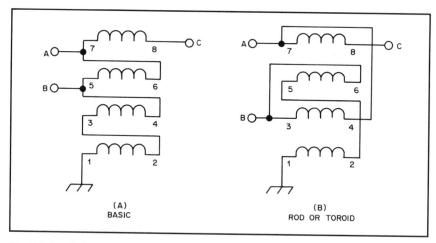

Fig 10-5—Schematic diagrams of quadrifilar windings for obtaining broadband ratios of 1:1.78 and 1:4. Schematic (A) is used for analysis purposes. Schematic (B) is used with both rods and toroids in matching 50-Ω coax at terminal C to 28 Ω at terminal A or 12.5 Ω at terminal B.

A) *Left:*

If an input voltage V_1 is connected from terminal B to ground, the bottom two transmission lines have $V_1 / 2$ on their inputs. Terminals 2 and 4 are "boot-strapped" by a potential of $V_1 / 2$ by their connections to terminals 3 and 5, respectively. Terminals 6 and 8 each have an added voltage of $V_1 / 2$ as a result of a voltage of $V_1 / 2$ traversing transmission lines. The added $V_1 / 2$ at terminal 8 traverses the transmission line twice since terminal 7 is connected to terminal 6. Thus, the voltage at terminal C then becomes $2V_1$, resulting in an impedance ratio of 1:4. Since the $V_1 / 2$ voltage from the top transmission line has twice the delay of the middle transmission line, the high-frequency response with the 1:4 ratio is not as good as the 1:1.78 ratio. By using small, high-permeability toroids (250 to 300), it is possible to cover 1.5 MHz to 30 MHz with the 1:4 ratio and still handle 1 kW of continuous power.

B) *Right:*

If an input voltage V_1 is connected from terminal A to ground, the three transmission lines have $V_1 / 3$ on their inputs. Terminals 2, 4 and 6 are then "boot-strapped" by a potential of $V_1 / 3$ by their connections to the odd-numbered terminals on the left. The voltage between terminals 8 and 6 is from the input $V_1 / 3$ which traverses the top transmission line, and thus the voltage at C becomes $4/3V_1$. Hence, the impedance ratio becomes $(V_{out} / V_1)^2 = (4 / 3)^2 = 1.78$. The top winding 7-8 carries three-fourths of the current into terminal A, and the other three only one-fourth of the current.

Fig 10-6 shows two versions of dual-output transformers with ratios of 1:1.78 and 1:4. Both of these transformers are capable of handling 1 kW of continuous power with either of their ratios. The chassis connectors are on the 50-Ω side (on the right side in Fig 10-5B). The parameters for the two transformers in Fig 10-6 are as follows:

A) *Left:*

8 quadrifilar turns on a $\frac{1}{2}$-inch diameter, 3.75-inch long rod ($\mu = 125$). Windings 7-8 and 3-4 are no. 14 wire. The other two are no. 16 wire. At the 1:4 ratio (input at terminal B), the response is essentially flat from 1.5 MHz to 21 MHz when matching 12.5 Ω to 50 Ω. At 30 MHz, the ratio increases some and also becomes complex. The optimum response (which is not much better) occurs at the 15:60-Ω

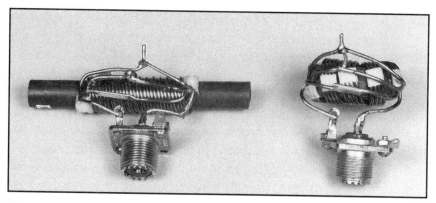

Fig 10-6—Two low-impedance, dual-output transformers with ratios of 1:1.78 and 1:4. Each use the schematic of Fig 10-5B.

level. At the 1:1.78 ratio (input at terminal A), the response is flat from 1.5 MHz to 30 MHz when matching 28 Ω to 50 Ω. This is also the optimum impedance level. This transformer again demonstrates the usefulness of rod transformers in Amateur Radio.

B) *Right:*

4 quadrifilar turns of no. 14 wire on a 1½-inch OD, K5 toroid (μ = 290). The turns are crowded to one side of the toroid in order to lower the characteristic impedance of the windings. At the 1:4 ratio (input at terminal B), the response is essentially flat from 1.5 MHz to 30 MHz when matching 12.5 Ω to 50 Ω. The optimum response occurs at the 15:60-Ω level, where it is flat from 1.5 MHz to 45 MHz. At the 1:1.78 ratio (input at terminal A), the response is flat from 1.5 MHz to 45 MHz when matching 28 Ω to 50 Ω. At the optimum impedance level of 34:60-Ω, the high-frequency response is well over 60 MHz. This is probably the best dual-output transformer, with ratios of 1:1.78 and 1:4, constructed by the author.

Sec 10.3 Parallel Transformers

As noted in the introduction to this chapter, two transmission line transformers can be connected in parallel on their input sides (usually the 50-Ω sides) and still offer their broadband ratios. The transformer that sees its proper match takes the load while the other one is essentially transparent. The author has constructed step-down, parallel-connected transformers matching a single 50-Ω coax to the various lower impedances of vertical, sloper and inverted-L antennas. This technique is akin to feeding parallel-connected dipoles (on separate bands) with a single coax. The small interaction between the transformers is due to the parasitic capacitance (that of a short transmission line) of the "floating" transformer. Since toroidal transformers require fewer turns, and hence have shorter transmission lines, their parasitic capacitances are lower than those of rod transformers. But at the 50-Ω input level, rod transformers have also been found to be acceptable in the frequency range of 1.5 MHz to 30 MHz.

This technique of parallel-connected transformers should also find use with other antenna systems. For example, a step-up transformer and a step-down transformer (from 50 Ω) should also work as well. By the use of 1:1 baluns on the outputs, beams and dipoles can also be fed from a single coax. Further, two antennas designed for the same frequency, but with different impedances, can also be matched. In this case, the input impedance of the parallel-connected transformers becomes 25 Ω. A series 2:1 unun can then bring the impedance back to 50 Ω.

It was also noted in the introduction that parallel-connected transformers have been used with various ground-fed antennas over a low-loss radial system (100 radials of 50 to 60 feet in length). The resonant antenna impedances seen by the transformers were between 12 and 35 Ω. If fewer radials are used, the added loss due to a poorer ground system has to be taken into account. As a reminder, information on loss v the number of ¼-λ ground radials is provided in Table 10-1. Also presented in the table is the resonant input impedance of a ¼-λ vertical

Table 10-1

Input Impedance and Ground Loss of a Resonant ¼-λ Vertical Antenna with Number of ¼-λ Radials

No. of Radials	Input Impedance (Ω)	Ground Loss (Ω)
1 (or ground rod)	85	50
4	65	30
8	57	22
40	39	4
100	36	1

with the added loss. It is safe to assume that the loss figures should also apply to ground-fed sloper and inverted-L antennas. As can be seen from Table 10-1, the multimatch transformers described in this chapter should be mainly used with ground systems of 40 or more $\frac{1}{4}$-λ (or longer) radials. On 160 meters, the 50- to 60-foot radials (100 of them) added only a few ohms of extra loss.

An example of transformers operating in parallel is shown in Fig 10-7, which is a photograph of two broadband, single-ratio transformers, connected in parallel on their 50-Ω (high) sides and mounted in a minibox. The parameters for these two transformers, which are capable of handling 1 kW of continuous power, are as follows:

A) *Left:*

This 1:1.56 ratio transformer has 5 quintufilar turns on a $1\frac{3}{4}$-inch OD, K5 toroid ($\mu = 290$). The schematic is shown in Fig 7-8B in Chapter 7. The center winding 5-6 is no. 14 wire. The other four are no.

Fig 10-7—An example of two single-ratio transformers operating in parallel. The transformer on the left has a 1:1.56 ratio, the one on the right, a 1:2.25 ratio.

16 wire. At the 32:50-Ω level, which is optimum, the response is flat from 1.5 MHz to well beyond 30 MHz.

B) *Right:*

This 1:2.25 ratio transformer has 8 trifilar turns on a 1½-inch OD, 4C4 toroid (μ = 125). The schematic is shown in Fig 7-20 in Chapter 7. The center winding 3-4 is no. 14 wire. The other two are no. 16 wire. At the 22.22:50-Ω level, which is optimum, the response is flat from 1.5 MHz to well beyond 30 MHz.

Sec 10.4   Dual-Output, Parallel Transformers

Fig 10-8 is an example of two broadband, dual-ratio transformers connected in parallel on their high-impedance sides, giving four broadband ratios for matching 50-Ω coax to lower impedances. The parameters for these transformers, which are capable of handling 1 kW of continuous power, are as follows:

Fig 10-8—An example of two broadband, dual-ratio transformers connected in parallel on their high-impedance sides, giving four broadband ratios for matching 50-Ω coax to lower impedances.

A) *Left:*

This transformer has ratios of 1:1.78 and 1:4. It has 4 quadrifilar turns on a 1½-inch OD, no. 64 toroid (μ = 250). The schematic is shown in Fig 10-5B. Windings 7-8 and 3-4 are no. 14 wire. The other two are no. 16 wire. As noted before, the windings are crowded to one side of the toroid in order to lower the characteristic impedances. At the 28:50-Ω level (using terminal A, the 1:1.78 ratio), the response is flat from 1.5 MHz to 45 MHz. This is also the optimum impedance level. At the 12.5:50-Ω level (using terminal B, the 1:4 ratio), the response is flat from 1.5 MHz to 30 MHz. The high-frequency response is improved somewhat at the 16.25:65-Ω level.

B) *Right:*

This transformer has ratios of 1:1.56 and 1:2.78. It has 4 quintufilar turns on a 1½-inch OD, no. 64 toroid (μ = 250). The schematic is shown in Fig 10-1B. The windings 7-8 and 3-4 are no. 14 wire. The other three are no. 16 wire. Winding 7-8 has two layers of Scotch no. 92 tape. At the 32:50-Ω level (using terminal A, the 1:1.56 ratio), the response is flat from 1.5 MHz to over 45 MHz. This is also the optimum impedance level. At the 18:50-Ω level (using terminal B, the 1:2.78 ratio), the response is flat from 1.5 MHz to 30 MHz. At the optimum impedance level of 23.5:65-Ω, the response is flat from 1.5 MHz to 45 MHz.

Sec 10.5   Eight-Ratio Transformer

Earlier work by the author produced an unbalanced-to-unbalanced transformer with eight separate ratios using a single toroidal core; it was shown in the first edition of this book. Although an improved version is possible (and will be discussed later in this section), the earlier version is reproduced here because it uses a popular toroid which is probably available in many radio amateurs' junk boxes. Fig 10-9 is a photograph of the transformer, Fig 10-10 is the schematic and Table 10-2 lists the performance at various impedance ratios using 50-Ω and 100-Ω load resistors. This transformer has 6 quadrifilar turns of no. 14 wire on a 2.4-inch OD, Q1 toroid (μ = 125). As shown in Fig 10-10, taps were at two turns from terminal 5 (F) and at five turns from terminal 5 (E). The useful frequency range in Table 10-2 is defined as the range where the loss is less than 0.4 dB. This loss at the high-frequency end is due to the transformation ratio becoming complex and increasing or

Fig 10-9—An 8-ratio quadrifilar transformer. Table 10-2 lists its
performance characteristics.

decreasing, depending upon the relationship between the load and the
effective characteristic impedance of the windings. Thus, the 0.4-dB
limit at the high end approximates a VSWR of 2:1 (equivalent to a
reflected power of 10%). This loss at the high end is not an ohmic loss
as such, but is due to the inability of the load to absorb the full available
power. Therefore, the useful ranges listed in Table 10-2 are very different from the ranges quoted on practically all of the other transformers
in this book. For these, the expressions "a flat response" or a "constant
impedance ratio" were used between a lower- and upper-frequency
limit, which amounts to a VSWR of 1:1 practically across the entire
frequency range.

Several improvements can be made in the 8-ratio transformer in
Fig 10-9. A smaller and higher-permeability toroid is recommended—
toroids with outside diameters of 1½ to 1¾ inches and permeabilities
of 250 to 300. Because of these changes, fewer turns are needed for
adequate low-frequency responses. Thus, 4 or 5 quadrifilar turns would

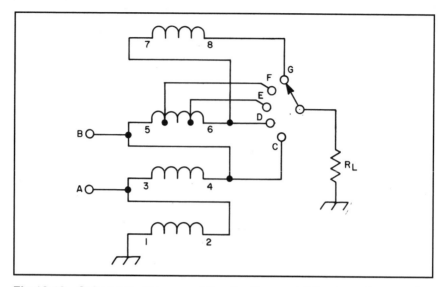

Fig 10-10—Schematic diagram of the 8-ratio quadrifilar transformer of Fig 10-9.

be sufficient. Fewer turns (and hence shorter transmission lines) increase the high-frequency performance. With fewer turns, the taps would also have to be changed. For the 4 quadrifilar-turns version, the taps should be set at one and three turns from terminal 5 (F and E), respectively. For the 5 quadrifilar-turns version, the taps should be set at one and four turns from terminal 5 (F and E), respectively. The resulting ratios would not differ greatly from those in Table 10-2. And finally, if this transformer is to be used for any length of time in matching 50 Ω to 3.125 Ω (16:1) at the kW level, then winding 1-2 (in Fig 10-10) should be replaced with no. 12 or even no. 10 wire. At this power level, the current in winding 1-2 becomes very high, since it handles three times more current than the 50-Ω coax. Further, this current tends to crowd between windings 1-2 and 3-4, where the electric field is a maximum.

Table 10-2

Performance of Quadrifilar Transformer

Input Port	Output Port	Impedance Ratio	Useful Frequency Range	
			R_L = 50 ohms	R_L = 100 ohms
B	F	1:1.36	1.5 to 30 MHz	3 to 30 MHz
B	E	1:2.0	1.5 to 30 MHz	3 to 30 MHz
B	D	1:2.25	1.5 to 30 MHz	3 to 30 MHz
A	C	1:4.0	1 to 30 MHz	3 to 30 MHz
A	F	1:5.4	1 to 15 MHz	1.5 to 30 MHz
A	E	1:8.0	1 to 15 MHz	1 to 15 MHz
A	D	1:9.0	1 to 15 MHz	1 to 30 MHz
A	G	1:16	1 to 8 MHz	1 to 15 MHz

Chapter 11

Materials and Power Ratings

Sec 11.1   Introduction

In Chapter 2 it was shown that the role of a transmission line transformer core is to enhance the choking action of the coiled windings to better isolate the input from the output circuit. Even though calculations were made for only the low-frequency performance of these transformers, experimental results demonstrated that the core influenced most of the operating region (see Sec 3.4).

The analysis for toroidal cores confirms that the low-frequency response is directly related to the permeability of the core material. With higher permeabilities, fewer turns are required, allowing for higher-frequency operation. It was also shown that rods, although not yielding the high reactance of the toroids, exhibit good low-frequency response and can be used in many applications. At low frequencies, the rod was found to be relatively insensitive to the permeability of the rod material because of the high reluctance of the air path around the core.

Accurate loss measurements have shown that only a limited number of ferrite materials are useful in power applications, where high efficiency is an important consideration. This chapter describes transmission line transformers that use nickel-zinc ferrite cores with permeabilities in the moderately low range of approximately 50 to 300, to yield efficiencies in excess of 98%. No conventional transformer can approach this performance. The losses are not a function of current as in the conventional transformer, but are, in most cases, related to the impedance levels at which the transformers are operated. This suggests a dielectric loss, rather than the conventional magnetic loss caused by core flux.

Included in this chapter are: (A) a history of ferrites, (B) experimental results on many of the typical ferrites when used with Ruthroff

1:4 unbalanced-to-unbalanced transformers, (C) new concepts in power ratings, and (D) suppliers of materials used in transmission line transformers. Although no efficiency measurements have been obtained on Guanella transformers, it can be safely assumed that the results would be as good as those found on Ruthroff transformers.

Sec 11.2 History of Ferrites

With the discovery of magnetic ferrites by T. Takei in Japan, and the excitement brought about by the 1947 publication of work done at the Philips Research Laboratories in the Netherlands during World War II, new and improved devices emerged in the field of magnetics (refs 30, 31).[1] The chemical formula for ferrites is MFe_2O_4, where M stands for any of the divalent ions: magnesium, zinc, copper, nickel, iron, cobalt or manganese, or a mixture of these ions. Except for compounds containing divalent iron ions, ferrites can be made with bulk resistivities in the range of 10^2 to 10^9 Ω-cm, compared to 10^{-5} Ω-cm for the ferromagnetic metals (such as powdered iron). This increase in bulk resistivity represents a major step forward for applications in frequency ranges heretofore unobtainable.

Ferrite compositions are made by ceramic technology. This involves intimate mixing of fine powders of appropriate oxides, compressing the mixture and firing it in carefully controlled atmospheres at temperatures of about 1100 °C to 1200 °C. Single crystals have been made by several techniques. By using different combinations of oxides previously listed and variations in ceramic processing, the mixtures can be tailored to fit a wide variety of technical requirements. In fact, ferrites with similar specifications by various manufacturers have been found to exhibit different efficiencies in transmission line transformers. This is because the ferrite is not completely defined by its chemistry and crystal structure; it is also defined by its processing, that is, powder preparation, compact formation, sintering and machining the ferrite to its final shape.

By the early 1950s, it was generally recognized that inductor cores of Permalloy dust had reached the point of diminishing returns in their application to higher and higher frequencies. A major contribution to the solution of the problem was made in 1952 by F. J. Schnettler and A. G. Ganz of Bell Labs, who developed a high-permeability manga-

---

[1]Each reference in this chapter can be found in Chapter 15.

nese-zinc ferrite for use at telephone carrier frequencies of 100 kHz and higher (ref 32). This material has also found widespread use at lower frequencies in power transformers, flyback transformers and deflection yokes.

In the 1960s and 1970s, Bell Labs scientists also made several important advances in linear ferrite properties, in response to the rising need for high-quality linear devices in the transmission area. This resulted in the development of a process for making suitable nickel-zinc ferrites capable of operating up to 500 MHz (ref 33). These advances were made by using cobalt additives and carefully controlled cooling. This form of ferrite is the best one for high-power transmission line transformers, and it is commercially available.

While the use of ferrites in inductors and transformers for carrier frequencies had a major impact on communications, its impact on microwave and computer technology is of equal importance. The availability of magnetic oxides eventually led to the large family of non-reciprocal magnetic devices that play a key role in microwave technology. The materials effort is credited largely to L. G. Van Uitert of Bell Labs. He proposed the substitution of nonmagnetic ions for magnetic ions in the ferrite structure to reduce internal fields and thereby lower the ferromagnetic resonance frequency.

In the computer field, A. Schonberg of Steatit-Magnesia AG in Germany, and workers at MIT's Lincoln Laboratories, found a family of magnesium-manganese ferrites with remarkably square hysteresis loops for use in memory and other computer and switching applications. These devices subsequently gave way to the semiconductor logic and memory circuits of the mid-1970s.

Although considerable information is available on the theory and application of transmission line transformers, dating back to the classic papers of Guanella in 1944 and Ruthroff in 1959, virtually no investigations have been made on the use of ferrites in power applications (refs 1, 9). Discussions by the author with scientists and engineers from major laboratories working in the ferrite field confirm this lack of development. The following sections will show the results obtained by the author on readily available ferrites and their use in high-power transmission line transformers. It is of some interest to note the differences in the properties of ferrites resulting from variations in processing techniques used by the different manufacturers.

Sec 11.3   Experimental Results

Early experiments by the author indicated that the bulk resistivity of ferrite material could be related to high-efficiency operation. Therefore, many of the major suppliers were asked to supply samples of their highest-resistivity material. Table 11-1 is a list of suppliers who provided samples, the code symbol for their materials, the low-frequency (initial) permeabilities and the bulk resistivities. Powdered iron was included because it has been used for some applications, but as will be shown, it suffers by comparison because of its very low permeability.

All of the data presented in this book on loss as a function of frequency were obtained at Bell Labs on a computer-operated transmission measuring set with an accuracy of one to two millidecibels over a frequency range of 50 Hz to 1000 MHz (refs 34, 35, 36). As a reference, it should be noted that a loss of 44 millidecibels represents a loss of 1%, or an efficiency of 99%. Actual data shows that many of the transformers, made using the best ferrite materials, exhibit losses over a consid-

Table 11-1

Cores, Suppliers and Specifications

Material	Supplier	Permeability	Bulk Resistivity $(\Omega\text{-}cm)$
Q1 (NiZn)	Allen-Bradley (formerly Indiana General)	125	10^8
G (NiZn)	Allen-Bradley	300	10^6
Q2 (NiZn)	Allen-Bradley	40	10^9
H (NiZn)	Allen-Bradley	850	$10^4 - 10^5$
4C4 (NiZn)	Ferroxcube	125	$10^7 - 10^8$
3C8 (MnZn)	Ferroxcube	2700	$10^2 - 10^3$
K5 (NiZn)	MH&W Intl (TDK)	290	2×10^8
KR6 (NiZn)	MH&W Intl (TDK)	2000	$10^5 - 10^6$
CMD5005 (NiZn)	Ceramic Magnetics	1400	7×10^9
C2025 (NiZn)	Ceramic Magnetics	175	5×10^6
CN20 (NiZn)	Ceramic Magnetics	800	10^6
C2050 (NiZn)	Ceramic Magnetics	100	3×10^7
E (Powdered Iron)	Arnold Engineering, Amidon Associates	10	10^{-2}

erable portion of their passbands of only 20 millidecibels, equivalent to efficiencies of 99.5%. Since short windings (wire lengths of 10 to 15 inches) are generally used, very little loss is attributed to the windings. In fact, wires as thin as no. 18 can easily handle a kW of power. Although the previous chapters stress theory and design, many of the experimental results presented do display the high efficiency of the transmission line transformer. This section reveals the differences between the various ferrites and, in particular, stresses the effect of operation at impedance levels greater than 100 Ω.

Figs 11-1, 11-2 and 11-3 show the loss results of three 4:1 transformers operating at three different impedance levels. Two are transmission line transformers and one is an autotransformer. The cores are similar in size, varying from 2.4 inches in OD for the Q1 cores and 2.625 inches for the K5 core. The two transmission line transformers had 5 tightly wound bifilar turns of no. 14 wire, and the autotransformer had 10 tightly wound turns of no. 14 wire. Fig 11-4 depicts this type of winding. Fig 11-1 shows that the two transmission line transformers had about the same loss above 2 MHz. Below 2 MHz, the K5 material was

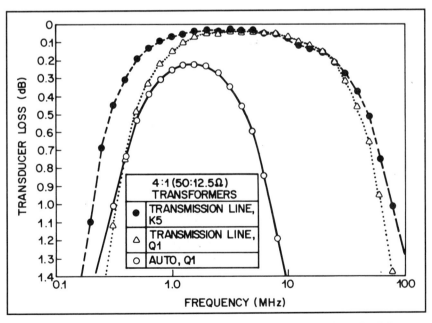

Fig 11-1—Performance of 4:1 transformers operating at the 50:12.5-Ω level.

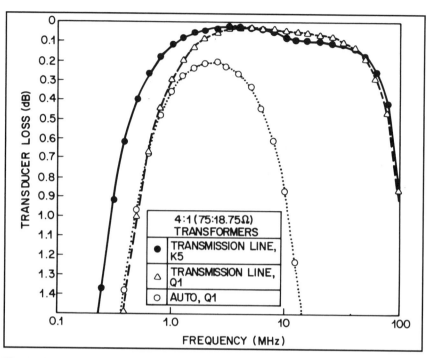

Fig 11-2—Performance of 4:1 transformers operating at the 75:18.75-Ω level.

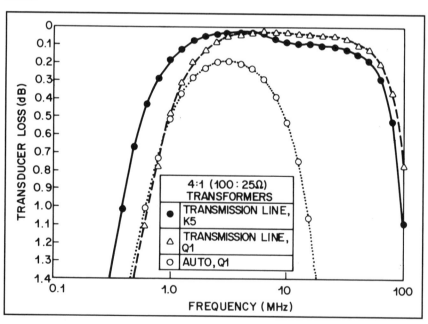

Fig 11-3—Performance of 4:1 transformers operating at the 100:25-Ω level.

superior because of its higher permeability (290 compared to 125 for Q1 material). The autotransformer showed not only a much narrower bandwidth, but also the greater loss of the conventional transformer (0.2 dB), which transmits the energy by coupling through flux linkages. Figs 11-2 and 11-3 show an interesting phenomenon of the K5 material. At about 7 MHz, another loss mechanism comes into play. At the higher impedance levels of 100:25 Ω of Fig 11-3, the loss beyond 10 MHz is more than double that of the Q1 material. In fact, at 10 MHz, the loss is only 35 millidecibels for the Q1 material and 80 millidecibels for the K5 material. Fig 11-3 also shows that the 100:25-Ω impedance level yields the better high-frequency response for the type of winding used in the test. Although it is difficult to see from the idealized drawings in these figures, actual data shows that the midband loss of the autotransformer decreases from 0.22 dB to 0.19 dB in going to the higher impedance level of 100:25 Ω. This confirms the expected decrease in core loss at the higher impedance levels, where the currents are lower.

To examine further the increase in the loss of transmission line transformers as a function of impedance levels, four 4:1 transformers, using different core materials with windings similar to Fig 11-4, were

Fig 11-4—The type of winding used in the 4:1 transformers of Figs 11-1, 11-2 and 11-3.

constructed and tested at the 200:50-Ω level. The transformers used K5, Q1, Q2 and Carbonyl E materials. The Q2-core transformer has 9 bifilar turns instead of 6 because of its lower permeability of 40. The results of the experiment are shown in Fig 11-5. The K5 and Q1 cores show more loss at the higher impedance level. Also, there is a considerable slope in the passband for these two materials, which shows that the loss is frequency dependent. But the Q2 material exhibited the same constant low-loss characteristic of the Q1 material at the lower impedance levels—about 40 millidecibels. The Carbonyl E core again exhibited a much poorer low-frequency response because of its low permeability of 10. Note that although the core's percentage bandwidth was small, the powdered iron exhibited a midband loss of only 0.07 dB at 50 MHz. This is still less than most of the ferrites listed in Table 11-1 operating at the 200:50-Ω level. And finally, the Carbonyl E core exhibited the highest frequency response because its core had a smaller cross sectional area, allowing a shorter transmission line with 6 turns. Conversely, the high-frequency response of the Q2 core was the poorest because of its much longer transmission line (9 turns).

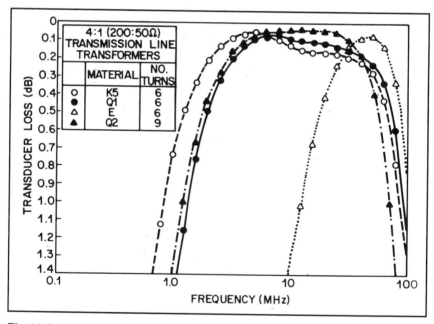

Fig 11-5—Loss v frequency of four 4:1 transformers at the 200:50-Ω level.

An interesting observation can be made on an earlier experiment which was described in Chapter 3. Fig 3-7 shows the performance of two transformers with optimized windings operating at the 40:10-Ω and 200:50-Ω level, each using 4C4 material. Although the higher-impedance transformer exhibited the better high-frequency response, which was the point of the experiment, the transformer also maintained about the same low-loss characteristic of the lower-impedance transformer. In other words, 4C4 material appears to be less sensitive to impedance levels. This difference between Q1 and 4C4 material may be caused by differences in processing. The 4C4 material is known to exhibit the normal B-H loop, while the Q1 material exhibits the perminvar loop, which has steeper sides and a smaller enclosed area.

Chapter 3 also shows data on G material (Fig 3-6). Losses of about 0.1 dB were noted at impedance levels of 75:18.75 Ω. This is about three times the loss of the Q1 and 4C4 materials. The permeability of the G material is also about three times larger than the other two.

Four ferrite toroids from Ceramic Magnetics were also investigated. The results are shown in Fig 11-6. These transformers used 6

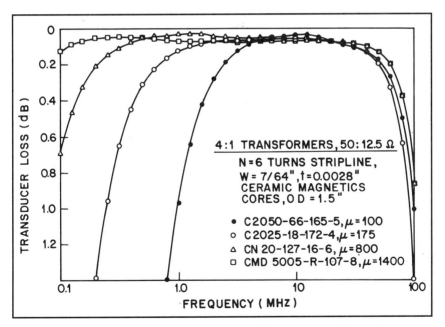

Fig 11-6—Loss v frequency for four core materials from Ceramic Magnetics with optimized windings for the 50:12.5-Ω level.

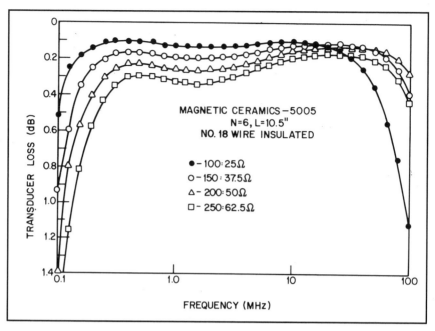

Fig 11-7—Loss v frequency for a CMD5005 toroid from Ceramic Magnetics with optimized winding for the 200:50-Ω level.

bifilar turns of 7/64-inch stripline, with Scotch no. 92 insulation, which is optimum for operation at the 50:12.5-Ω level. The cores had an OD of 1½ inches. Although their permeabilities ranged from 100 to 1400, the losses were about the same for the four different cores—less than 0.08 dB over their passbands. This is particularly noteworthy for the CMD5005 material, which has a permeability of 1400. It is the lowest loss measured by the author on any of the higher-permeability materials. Coincidentally, its bulk resistivity of 7×10^9 Ω-cm is the highest of any of the ferrites. Fig 11-7 shows the results of the CMD5005 material at higher impedance levels. This transformer used 6 bifilar turns of Teflon-covered no. 18 wire, optimized for the 200:50-Ω level. Note that the CMD5005 material shows an increased loss similar to that which most ferrites exhibit at higher impedance levels.

These measurements on Ceramic Magnetics materials also bring out another important fact about transmission line transformers. Conventional wisdom has stated that increased coupling between transmission lines is necessary for better low-frequency response. The results in Fig 11-6 show that the differences in low-frequency responses between these four materials are due to the differences in their permeabilities.

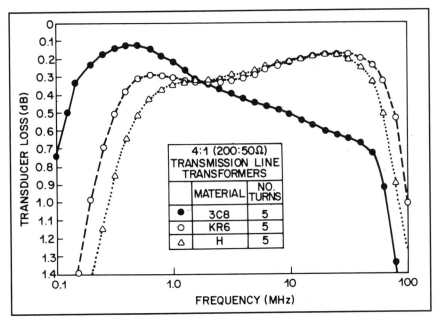

Fig 11-8—Measurements of 3C8, KR6 and H materials at the 200:50-Ω level.

The transmission lines, and hence couplings, are the same. What is taking place here is that the reactance of the windings (and hence choking action) increases with permeability, thus resulting in better low-frequency response.

Finally, Fig 11-8 shows measurements taken on 3C8, KR6 and H materials at the 200:50-Ω level. Even at lower impedance levels, these materials exhibit considerably more loss than the other materials already described. Incidentally, the 3C8 material is a manganese-zinc ferrite and has a high permeability of 2700. It is used extensively at low frequencies in power transformers, TV flyback transformers and deflection yokes. Ruthroff used this type of ferrite in his original work on transmission line transformers, but it is unsuitable for applications at power levels, where efficiency becomes a factor (ref 9).

Sec 11.4   Power Ratings

Power ratings are generally determined by two conditions: (1) the temperature rise due to losses, and (2) the exceeding of some maximum value of operating parameter by accident, which can create a catastrophic failure. A failure caused by an increase in temperature is usually time dependent, while the breakdown of a device operated over its preset value is instantaneous.

With conventional transformers, losses can be separated into three categories: eddy-current losses, hysteresis losses and residual losses. These losses, together with the permeability of ferrite material, generally increase with temperature to create a possible runaway condition. Catastrophic failures can occur with some ferrite materials at flux densities greater than 500 gauss. This is particularly true of ferrites exhibiting the perminvar loop; they become excessively lossy. Each of these conditions are related to flux densities and hence ampere-turns. Increasing the size of the core can, therefore, increase the power rating.

Different conditions exist when a transmission line transformer's power rating is determined. Because of the canceling effect of the transmission line currents, little flux is generated in the core. This holds true even when tapped multiwindings are involved. Since losses with certain ferrites are only on the order of 20 to 40 millidecibels, very small transmission line transformers can handle surprisingly high power levels. Further, as was shown in the previous section, losses in most ferrite materials increase with impedance levels, and these levels must be considered when designing a transmission line transformer. Most failures in transmission line transformers are of the catastrophic type, and are usually caused by poorly terminated (or unterminated) transformers. Such conditions create high voltages and a breakdown in the insulation between the windings. This is particularly true of close-wound enameled or Formvar-type wires.

Standards for setting power ratings for transmission line transformers have not appeared in the literature, nor are they available from any of the suppliers of ferrite material. To the author's knowledge, the data presented in this book is the only quantitative information available on the losses of these transformers. Because limited reliability information is available on transmission line transformers, and losses generally increase with impedance levels, an exact formulation of power ratings is difficult to make as of this writing. But as a result of the findings by the author, some general guidelines can be offered when considering ratings. They are:

1) The power capability of these devices (when energy is transmitted from input to output by a transmission line mode) is determined more by the size of the conductors, and not by the cores. Very small structures can handle amazingly high power levels. Thus, larger wires

or the use of coaxial cable or stripline can more than double the power ratings.

2) The voltage ratings can be increased significantly by the use of polyimide-coated wires. Some commercial brands are ML and H Imideze wires. In many cases, in order to optimize the characteristic impedance of the windings, extra layers of Scotch no. 92 tape (another polyimide insulation) are used. This also increases the breakdown voltage.

3) Generally, the lowest-permeability nickel-zinc ferrites yield the highest efficiencies. These have permeabilities in the range of 40 to 50. But, these ferrites can limit the low-frequency response. When operating at impedance levels below 100 Ω, permeabilities as high as 300 should yield very high efficiencies (98% to 99%). When operating at impedances above 100 Ω, the trade-off is low-frequency response for efficiency. Actually, most of the ferrites with permeabilities of 200 to 300 can still yield acceptable efficiencies (at least 97%) at the 200- to 300-Ω impedance level.

4) Very few differences in efficiency were observed from the ferrites supplied by the manufacturers listed in Sec 11.5. Limited measurements on 4C4 material ($\mu = 125$) from Ferroxcube showed the best efficiency at the 200-Ω level. This material is also reported to be free of the failure mechanism due to high flux density (Sec 11.3) exhibited by most of the other ferrites.

5) Although many examples in this text refer to the various company designations for the ferrites used, practically any other ferrite (with the same permeability) can be substituted for them.

6) When transformers become warm to the touch (after the power is turned off!), it suggests that either the wrong ferrite is used or that the reactance of the coiled windings, at the frequency in question, is insufficient to prevent conventional transformer currents. The problem is probably not in the size of the conductors.

Table 11-2 lists some suggested power ratings. Although efficiencies can vary with permeability and impedance level, these ratings should generally hold for permeabilities below 300. Also, since the currents can vary with impedance level and the position of the winding in a higher-order winding, the transmission line descriptions are only offered as a general practice.

Several small transformers were tested under severe conditions to check the validity of the ratings in Table 11-2. One 4:1 transformer

Table 11-2

Suggested Power Ratings for Ferrites with Permeabilities Below 300

Core Size	Description of Transmission Line	Rating (Continuous Power)
1-inch OD toroid, ¼-inch diameter rod	16-18 gauge wire	200 W
1½-inch OD (or greater) toroid, ⅜-inch diameter (or greater) rod	14-gauge wire	1000 W
1½-inch OD (or greater) toroid, ⅜-inch diameter (or greater) rod	10-12 gauge wire, coaxial cable, stripline	2000 W

had 10 turns of no. 18 wire on a Q1 toroid with a 1-inch OD. The other transformer had 14 turns of no. 18 wire on a Q1 rod with a ¼-inch diameter. These transformers, operating at an impedance level of 50:12.5 Ω, successfully handled 1 kW of peak power in single-sideband operation over an extended period of time. They became warm to the touch, but showed no evidence of damage.

Sec 11.5 Suppliers of Materials

Most of the feedback from the first edition of this book was concerned with the need for more practical designs and information on obtaining materials. Many new designs were added to this second edition as a response to the first concern. This section attempts to satisfy the second concern by listing suppliers in Table 11-3. Several comments can be made regarding this listing. Three general distributors handle practically all of the materials (or suitable substitutes) used in these devices and are usually listed in *QST*. Many of the manufacturers also have distributors and they are also listed here. A few manufacturers have discontinued their operation in recent years (like Indiana General) and, as time goes on, others could be added to or subtracted from this listing. Table 11-3 is a list of suppliers that the author has contacted and whose materials were used in the course of preparing the information presented in the first and second editions of this book.

Table 11-3 — Material Suppliers

General Distributors (all materials)

Amidon Associates
2216 E Gladwick St
Dominguez Hills, CA 90220
213-763-5770

Certified Communications
"The Wireman"
261 Pittman Rd
Landrum, SC 29356
800-727-WIRE

Palomar Engineers
Box 455
Escondido, CA 92025
619-747-3343

Ferrite Manufacturers

Ceramic Magnetics
87 Fairfield Rd
Fairfield, NJ 07006
201-227-4222
Ferrite: CMD5005

Fair-Rite Products Corp
Box J
Wallkill, NY 12589
914-895-2055
Ferrites: 61, 64, 66, 67

FDK
Fuji Electrochemical Co, Ltd
5-36-11, Shinbashi, Minato-ku,
Tokyo 105, Japan
Ferrites: H52A, H53Z
Distributor:
FDK America Inc
17326 Edwards Rd, A-130
Cerritos, CA 90701
213-404-1770

Philips Components (Ferroxcube)
5083 Kings Hwy
Saugerties, NY 12477
914-246-2811
Ferrite: 4C4
Distributor:
Elna Ferrite Laboratories, Inc
234 Tinker St
Woodstock, NY 12498
914-679-2497

Krystinel Corp
126-140 Pennsylvania Ave
Paterson, NJ 07509
201-345-8900
Ferrite: KM1
Distributor:
Permag Central Division
1050 Morse Ave
Elk Grove, IL 60007
708-956-1140

Magnetics
A Division of Spang and Company
900 E Butler Rd
Box 391
Butler, PA 16003
412-282-8282
Ferrite: A

Siemens
Special Products Division
186 Wood Ave S
Iselin, NJ 08830
201-321-3400
Ferrites: K1, M33
Distributor:
NECCO
2460 Lemoine Ave
Fort Lee, NJ 07024
201-461-2789

TDK Electronics Co, Ltd
14-6, 2-Chome, Uchikanda
Chiyoda-ku, Tokyo 101, Japan
Ferrite: K5
Distributor:
MH & W International Corp
14 Leighton Pl
Mahwah, NJ 07430
201-891-8800

Tokin Corp
Hazama Bldg
Ni-chome, Kita-aoyama
Minato-ku, Tokyo 107, Japan
Ferrite: 250L
Distributor:
Tokin America Inc
9935 Capitol Drive
Wheeling, IL 60090
708-215-8802

Wire and Insulation Manufacturers

Phelps Dodge
Magnet Wire Company
Box 755
Fort Wayne, IN 46801
800-255-2542
Wire: H Imideze

3M Company
Industrial Electrical Products Division
3130 Lexington Ave S
Eagan, MN 55121
800-233-3636
Insulation: Scotch nos. 27, 92

Chapter 12

Simple Test Equipment

Sec 12.1 Introduction

This chapter is directed to the person who does not have access to sophisticated test equipment and must rely on simple equipment which can be constructed from readily available parts. The material focuses on homemade test gear that can give surprisingly good results. Measurements made on this equipment checked very closely with high-precision laboratory test sets (refs 34, 35, 36).[1]

All of the important transformer characteristics can be measured using this basic apparatus. Parameters such as transformation ratios, high- and low-frequency performance, optimum impedance levels for the types of windings used, and the characteristic impedances of bifilar windings can be readily determined. Each parameter can be evaluated with a simple resistive bridge and a general-coverage signal source that uses junction field-effect transistors (JFETs). That elusive parameter known as efficiency can also be obtained satisfactorily by a simple comparative technique described in Sec 12.5. This is an indirect measure of the in-band loss (only 20 to 40 millidecibels). (Direct measurements can be made only with highly complex laboratory apparatus.) As an added bonus, this simple equipment can be effective for measuring the important parameters of vertical antennas, that is, the resonant frequency and the resistance at resonance.

Sec 12.2 The Wheatstone Bridge

The simple resistive bridge, known as the Wheatstone bridge, is shown in Fig 12-1A. The bridge is balanced when no voltage exists between terminals D and B. At this point of balance, no current flows

---

[1]Each reference in this chapter can be found in Chapter 15.

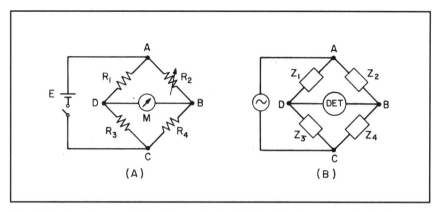

Fig 12-1—A schematic for the resistive bridge is shown at A; B shows an impedance bridge.

through the galvanometer (M) and a null is indicated. This occurs when the ratios in the arms are

R1 / R3 = R2 / R4   (Eq 12-1)

If R4 is the unknown resistance, then Eq 12-1 can be rewritten as

R4 = R2 × R3 / R1   (Eq 12-2)

In general, the ratio R3 / R1 is some conveniently fixed value such as 1, 2, 3, and so on, and R2 is made variable for balancing purposes. By using many known values of R4, the dial of the variable resistance, R2, can be readily calibrated. For example, if the dial has 100 divisions to cover one complete rotation of R2 (usually 330 degrees), and R1 = R3, then the dial can be read directly in ohms. If R3 / R1 = 2, then each division equals 2 Ω (or 200 Ω full scale). Many other combinations are possible. The simple ac impedance bridge shown in Fig 12-1B is an extension of the Wheatstone bridge. The bridge is balanced when the voltage between terminals D and B is zero. This occurs when

Z1 / Z3 = Z2 / Z4   (Eq 12-3)

The unknown impedance Z4 becomes

Z4 = Z2 × Z3 / Z1   (Eq 12-4)

If Z3 and Z1 are pure resistances, and Z4 is complex (that is, R4 + jX4), then Z2 must have a resistive and a reactive component for balance. Again, other combinations are possible. Ac bridges can be used to trade off capacitive reactances for inductive reactances by proper placement of the arms in the bridge.

Sec 12.3   A High-Frequency Resistive Bridge

Accurate ac bridges are capable of measuring reactance as well as resistance, but are rather difficult to construct and calibrate. Surprisingly, the simple resistive bridge, carefully constructed to reduce parasitics, can be used to measure the performance of broadband transformers. The bridge is effective up to 100 MHz or more. The trick (or more properly, the technique) is to understand the nature of the null. With pure resistive loads (R4), the nulls are very sharp and deep. This, then, becomes the standard indication for any purely resistive load. Within the transformer passbands, when they are terminated with resistive loads, their input impedances are also resistive. Thus, the nulls of the bridge are generally sharp and deep. To reduce the effect of inductive parasitics, it is advantageous to terminate the transformer in question with a resistive load on its high-impedance side and measure from the low-impedance side. Thus, within the passband of the transformers, transformation ratios are readily measured, as reactive components are not created. The devices act as ideal transformers.

The low-frequency response can be determined with the aid of a variable-frequency signal source. At the low-frequency end, when the choking action of the coiled transmission line becomes inadequate, the input impedance of the terminated transformer takes on an inductive component, causing the null to be less pronounced. By lowering the frequency of the signal source to where the meter reading at the null increases noticeably (usually 10% to 20% of full scale), a fairly good measure of the low-frequency capability is obtained.

In a similar fashion, the high-frequency response can be determined by increasing the frequency of the signal source until a noticeable increase in the null reading takes place. This indicates a falling-off of the transformer's capability at the high-frequency end. Also of interest here is that the optimum impedance level for maximum high-frequency response can be determined easily. By varying the value of the terminating resistance and noticing the depth of the null at the high-frequency end, the best impedance level for the type of winding used on the transformer can be determined. As will be shown in Sec 12.6, even the

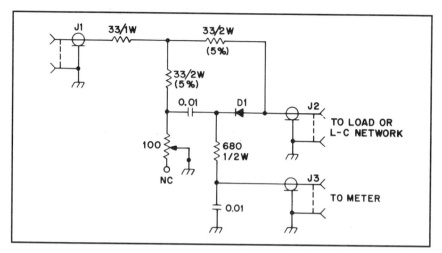

Fig 12-2—A schematic of the linear resistive bridge.

characteristic impedance of short (coiled or straight) transmission lines can be evaluated quite accurately by this method.

Fig 12-2 shows a circuit diagram for a linear resistive bridge. A germanium diode, D1, is used because of its lower forward-voltage drop as compared to a silicon diode. The fixed 33-Ω resistances in the top arms of the bridge are not especially critical. Values of 47 Ω to 68 Ω have also been used. Fig 12-3 shows the bottom view of a bridge and Fig 12-4 shows the top view. To calibrate the dial, various calibration

Fig 12-3—The bottom view of a linear resistive bridge.

Fig 12-4—The top view of a linear resistive bridge.

resistors can be connected from J2 (via a banana plug) to the grounded binding post. It is recommended that the calibration be done with 1% noninductive resistors (which are available at many local radio stores), and at 3.5 MHz or lower in frequency. In this frequency range, the inductive effect of the resistor leads is minimized. This bridge, with its short leads, works well up to at least 100 MHz, with an accuracy of about 1 Ω over its linear scale.

Fig 12-5 shows a circuit diagram for the current amplifier used in conjunction with the resistive bridge of Fig 12-2. With a 50-µA meter,

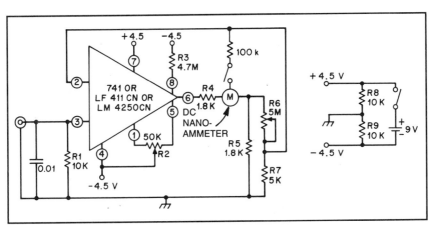

Fig 12-5—A schematic of the sensitive current amplifier used in conjunction with the resistive bridge.

the maximum sensitivity is 7 nA full scale. Several interesting features should be noted in Fig 12-5:

1) By placing R6, a 5-MΩ potentiometer, in the feedback arm, sensitivity may be controlled without causing an imbalance in the operational amplifier. Many op amps are sensitive to the gain controls on the input circuit.

2) Since op amps can now work at much lower voltages than before, the 10 kΩ—10 kΩ voltage divider circuits make 9-V battery operation practical.

3) The 100-kΩ resistor can be switched in parallel with the 5-MΩ potentiometer to act as a useful high-low sensitivity control.

Fig 12-6 shows the front view of a sensitive current amplifier. Fig 12-7 shows a top view of the resistive bridge and sensitive current amplifier within one enclosure. Fig 12-8 shows the bottom view. This instrument is particularly useful in conjunction with vertical antennas.

Sec 12.4   Signal Generators

When you measure the performance of broadband trans-formers (as well as vertical antennas), it is convenient to have signal sources that are continuously variable over a very large frequency range, compatible in signal level with the impedance bridges, constant in output and portable. The sources in this section fulfill these requirements.

Fig 12-9 is a schematic using JFETs in a Colpitts oscillator and a source follower. The transistors, Q1 and Q2, are J212 and J309, acquired from National Semiconductor. Both components are capable of operating beyond 100 MHz. J309 is particularly useful; its high transconductance of 20,000 microsiemens (µS) gives an output of

Fig 12-6—The front view of the sensitive current amplifier.

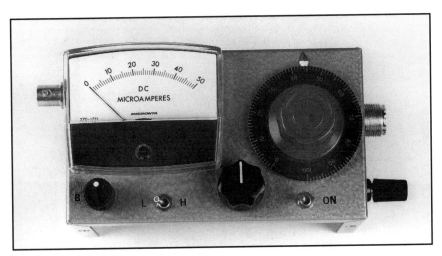

Fig 12-7—The top view of a resistive bridge and sensitive current amplifier in one enclosure.

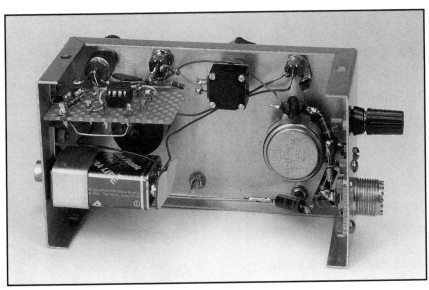

Fig 12-8—A bottom view of a resistive bridge and sensitive current amplifier in one enclosure.

50 Ω as a source follower. Its input impedance is also extremely high, thus making it ideal for use in a decoupling stage. Other JFETs have worked satisfactorily in the circuit of Fig 12-9: Motorola's 102, 105, 107 and Texas Instruments 2N5397. The inductors used are from the J.

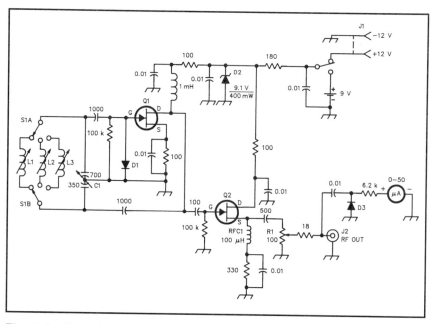

Fig 12-9—This signal source uses junction field-effect transistors as a Colpitts oscillator and source follower.

W. Miller Co and have the following numbers and values: L1 is a 4508 (24-35 µH), L2 is a 4503 (1.6-2.8 µH), and L3 is a 4501 (0.4-0.8 µH). Fig 12-10 is the signal generator's front view, and Fig 12-11 shows its rear view. The tuning dial was calibrated against a general-coverage receiver. A 12-inch length of wire is connected to the center conductor of the output jack to allow for sufficient receiver pickup. Fig 12-12 shows the dial calibration curves.

Fig 12-10—The front view of a signal source using a JFET Colpitts oscillator and source follower.

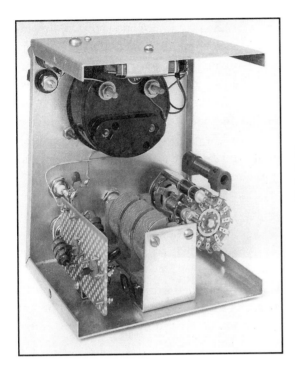

Fig 12-11—The rear view of the signal source shown in Fig 12-10.

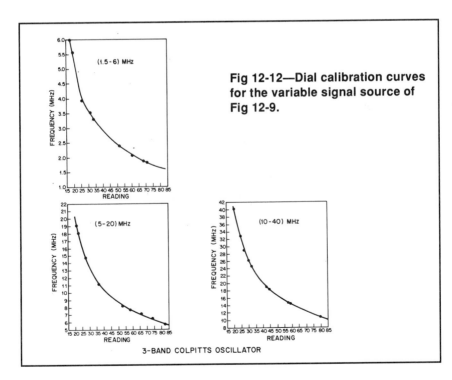

Fig 12-12—Dial calibration curves for the variable signal source of Fig 12-9.

Another generator using a Colpitts oscillator and source follower is shown in Fig 12-13. This unit uses a large National dial which can be calibrated directly on its face. This obviously offers the advantage of more easily determining the frequency of operation. The differences between the signal sources of Fig 12-13 and Fig 12-10 are in the sizes of the variable capacitors and the number of inductors. Fig 12-13 uses a 365-pF dual-gang variable instead of a 350-pF triple-gang variable. The inductors used in the unit of Fig 12-13 are also from the J. W. Miller Co and have the following numbers and values:

L1—4409 (68-130 μH)
L2—4408 (30-69 μH)
L3—4506 (9-16 μH)
L4—4503 (1.6-2.8 μH)
L5—4502 (1.0-1.6 μH)
L6—4501 (0.4-0.8 μH)

This signal source has a broader range; it operates from 1 to about 70 MHz with practically constant output. By removing half of the rotor

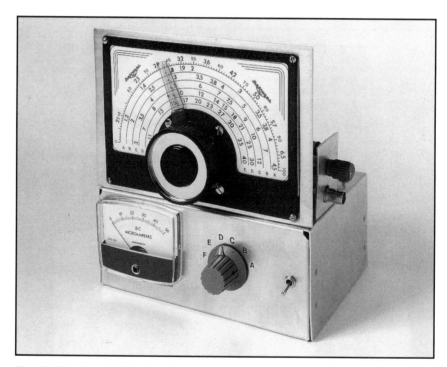

Fig 12-13—A variable signal source with dial calibration.

plates of the variable capacitor in Fig 12-13, the high-frequency range is increased to about 100 MHz. This procedure tends to spread out the tuning capability in the lower ranges, which is helpful in more accurately determining the frequency. Fig 12-14 shows the back view of this signal source, and Fig 12-15 shows the bottom view (with bottom plate removed).

Another generator using a Hartley oscillator and a source follower is shown in Fig 12-16. The Hartley enjoys the advantage of requiring only a single variable capacitor, but at the expense of using tapped coils and of isolating the variable capacitor from ground. All coils are homemade and tapped at about ¼ of the turns from the top (the smaller number of turns appears between the gate and the source of the JFET). The information on the four inductors is as follows:

L1—6 turns no. 18 wire on ⅜-inch ceramic form

L2—15 turns no. 18 wire on ⅜-inch ceramic form

L3—28 turns no. 20 wire on ½-inch ceramic form

L4—34 turns no. 20 wire on ½-inch ceramic form

Fig 12-17 is the top view of the generator, and Fig 12-18 is the bottom view. Fig 12-19 shows the calibration curves.

Fig 12-14—The back view of the variable source of Fig 12-13.

Fig 12-15—A bottom view of the variable source of Fig 12-13.

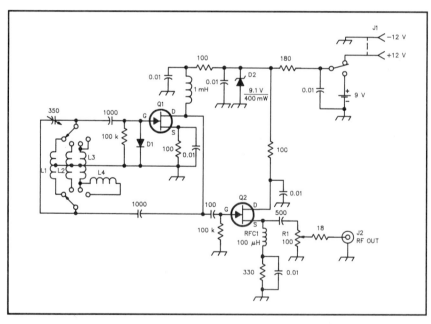

Fig 12-16—The schematic of a signal source using a Hartley oscillator and a source follower.

Fig 12-17—A front-panel view of the Hartley oscillator signal source.

Fig 12-18—A bottom view of the Hartley oscillator signal source.

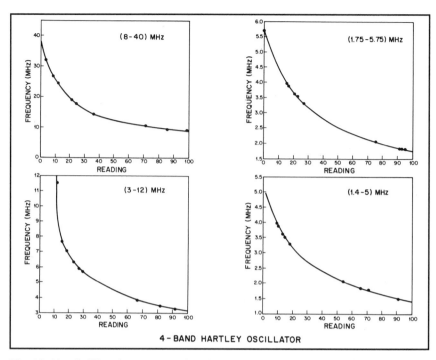

Fig 12-19—Calibration curves for the Hartley oscillator signal source.

Several comments can be made regarding these signal sources. When calibrating them (using about a 10-inch antenna on the output jack) with a general-coverage receiver, one finds the signals to vary somewhat and the tuning to be rather sharp. But these sources have been found to have more than adequate stability from 1 MHz to 30 MHz. This is because the bandwidths of transmission line transformers and ground-fed antennas (even on 160 meters) are relatively wide, and therefore stability is not an issue. The equivalent of J. W. Miller RF coils have been used in the signal sources shown in Figs 12-10 and 12-13. Another coil supplier is CTC. In any event, home-made coils using small plastic forms for the higher frequencies and small powdered-iron toroids (red mixture, $\mu = 10$) for the lower frequencies can be found to be adequate. Since these coils are not variable (they lack the adjustable powdered-iron slug), careful pruning should bring them into the proper ranges.

Sec 12.5   Efficiency Measurements—The Soak Test

As was mentioned in previous chapters, if proper ferrite materials are used for the cores of transmission line transformers, they can exhibit outstandingly high efficiencies. This occurs because of the canceling effect of the transmission line currents. Energy is transmitted by a transmission line mode instead of by flux linkages as in a conventional transformer. This holds true at the transformer's high-frequency limits, where standing waves come into play to create a complex transformation ratio which can be different from that of the mid-band ratio.

If an accurate gain and phase test set for determining efficiency is not accessible, a simple technique can be employed which gives surprisingly good results. This technique is called (by the author) a soak test. The test is able to distinguish transformers that are 95% efficient or less from those that are 98% or 99% efficient. The soak test involves using a known, efficient transformer in series with an unknown transformer. The transformers are essentially connected back to back and inserted in a coaxial cable line where appreciable power is transmitted. For example, a 4:1 step-down transformer (say 50 Ω to 12.5 Ω) would be in series with a 1:4 step-up transformer. A 50- to 200-Ω transformer could equally be in series with a 200- to 50-Ω transformer to obtain the original impedance of the coaxial cable. The same is true for baluns, fractional-ratio transformers and other devices with higher transformation ratios. Fig 12-20 shows two transformers connected in a back-to-back arrangement.

After power is applied for several minutes, touch the transformers (with the power off!) to see if a noticeable temperature rise has occurred. Transformers with proper ferrite cores and with no. 14 or 16 wire on toroids of 1½-inch OD or greater, or on rods of ½ inch in diameter, virtually have no detectable (by touch) temperature rise while handling 1 kW of single-sideband power. Transformers with efficiencies of 95% or less show noticeable temperature rises. Incidentally, the transformers of Fig 12-20 use no. 18 wire on the 1-inch OD toroid and the ¼-inch diameter rod (both of Q1 material). These transformers became only mildly warm and exhibited no permanent damage when operating at the 1-kW peak power level. The rod transformer was a little warmer than the toroid, because it had 40% more wire. At 200 W, neither transformer showed any perceptible temperature rise.

Fig 12-20—Two 4:1 transformers connected back-to-back for a soak test.

Sec 12.6   Characteristic Impedance Measurements

As was stated many times throughout this text, when adequate isolation (due to the coiling of transmission lines around a core or the threading through beads) exists between the input and output, these devices then transfer the energy by an efficient transmission line mode. Thus, their designs mainly depend upon transmission line theory and practice. As with conventional transmission lines, the characteristic impedance also plays a major role with transmission line transformers. Characteristic impedances appreciably greater or smaller than the optimum value can seriously affect the high-frequency response. The design goal is to be within 10% of the optimum value. For the 1:4 transformers, whether they be Ruthroff or Guanella designs, the optimum characteristic impedance is one-half the value of the resistance on the high-impedance side. For low-impedance coaxial cable or stripline transformers, the optimum characteristic impedance has been found (experimentally) to be some 80% to 90% of this value. Figs 4-3 and 5-2 show curves for the characteristic impedance of various kinds of transmission lines which were obtained with the simple resistive bridges described in Sec 12.3. These results were from measurements on transmission lines only 10 to 20 inches in length! The resistive bridge, with a very sensitive detection arrangement, is an excellent detector of phase (the depth of the null is diminished when the bridge sees a nonresistive termination). Thus, terminating these short transmission lines (straight or coiled) with various noninductive resistors until the depth of the null approaches that of a pure resistor (resulting in a "flat" line) gives a quick and accurate measurement of the characteristic impedance. With this method (which rivals that of any sophisticated impedance bridge), the accuracy is solely determined by the calibration of the resistive bridge or the true values of the terminating resistors.

Fig 12-21 shows two of the examples used in obtaining data for the characteristic impedance as a function of wire size and spacing (Fig 5-2). The 2.4-inch OD toroid on the left has 5 bifilar turns of no. 12 H Imideze wire which is covered with a 25-mil wall Teflon tubing. The separation between the wires (also considering the wire coating) is 60 mils and the characteristic impedance measured 97 Ω. The 1-inch OD toroid on the right has 12 bifilar turns of no. 22 hook-up wire. The insulation on the wires is 9 mils thick, and thus the separation between

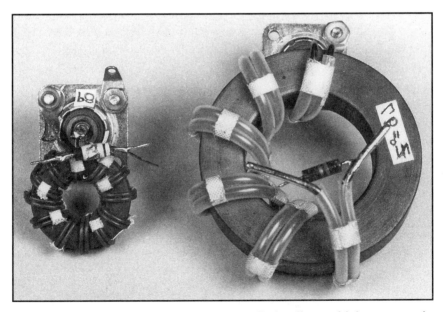

Fig 12-21—Two examples of coiled transmission lines which were used in obtaining data for the characteristic impedance as a function of wire size and spacing (Fig 5-2).

the wires is 18 mils. The characteristic impedance also measured 97 Ω. The following are comments and suggestions regarding measurements of characteristic impedances (especially with the equipment described in this chapter):

A) In measuring impedances between 90 and 250 Ω, the range of the bridge in Fig 12-2 has to be increased 2½ times. The author found that replacing the 100-Ω pot with a 250-Ω pot, and the two 33-Ω resistors (in the top arms) with 68-Ω resistors, results in an excellent resistive bridge for this range of impedances.

B) A straight, wire transmission line always yields a higher impedance value than a coiled one. This is due to the proximity effect of adjacent bifilar turns in a coiled transmission line. The smaller the spacing is between bifilar turns, the lower is the characteristic impedance.

C) A straight coaxial cable transmission line (of the low-impedance types described in this book) always yields a higher value than the one that is coiled around a core. This is because the effective spacing between the inner conductor and the outer braid decreases with coiling. The proximity of neighboring turns does not effect the characteristic impedance, but does effect the parasitic capacitance, and hence the high-frequency response.

D) The most accurate data was obtained between 10 and 20 MHz. In this frequency range, the phase angle can still be large (if the termination is not equal to Z_0) and easily detected. Further, the parasitic inductance of the leads of the terminating resistors is minimal.

E) For all forms of measurements dealing with transmission line transformers, the author found the best settings on the current amplifier to be 75% to 100% of full sensitivity, and on the signal source to be about 25% of full scale. At high frequencies, where considerable phase angle can exist, the sensitivity of the current amplifier has to be reduced appropriately.

F) For the most meaningful readings, the characteristic impedance measurements should be made in the transformer's final configuration. In this way, all interactions between the windings are taken into account.

Chapter 13

Hints and Kinks

Sec 13.1   Introduction

Many of the changes presented in the second edition of this book are a result of the large number of requests by radio amateurs for more practical designs and "how-to-do." Therefore, some 100 designs have been added to this second edition to help satisfy the first request. Certainly not all needs are answered, since the author's interest and experience has been mainly related to matching coaxial cable to ground-fed antennas in the 1.7-MHz to 30-MHz range. Some of the designs bordered on other areas of interest, such as hybrids, antenna tuners and the VHF and UHF bands. It is hoped that sufficient information on design principles is presented in this second edition to enable readers working in these other areas to design successful transformers.

This chapter is concerned with the "how-to-do"; how to: (A) select the proper ferrites, (B) wind rod and toroidal transformers, (C) construct low-impedance coaxial cable and (D) handle and take care of ferrite transformers. The techniques described in the following sections evolved over many years of winding hundreds of transformers (it takes the author about three attempts in order to arrive at the final design). It is quite certain that there are many other techniques that can do the job as well (or even better). But those shown in this chapter have worked well for the author, and are offered here for the experimentalist.

Sec 13.2   Selecting Ferrites—Substitutions

Efficiency measurements on many ferrites from major manufacturers have shown that the highest values have been obtained from nickel-zinc material with permeabilities less than 300. Improvements are continually being made in the manufacturing process, and this upper limit of 300 will probably increase in the future. Manganese-zinc ferrites, which have permeabilities even exceeding 10,000, are highly lossy when used as cores for transmission line transformers and there-fore are not recommended in power applications. For those who have obtained some unknown rods and/or toroids and want assistance in identifying them, here are some suggestions:

A) *Appearance:* Ferrites can come in all shades of black and gray-black. They can be either shiny or dull. In some cases they can have a protective coating (for the wire). Therefore, low-permeability nickel-zinc ferrite is indistinguishable in appearance from high-perme-ability nickel-zinc ferrite or from manganese-zinc ferrite. But pow-dered-iron toroids (which are not recommended because of their very low permeabilities) usually have distinctive protective coatings. The popular T-200-2 toroid has a clear enamel-like finish and a definite undercoating of red. It has a permeability of only 10 and is called the "red mixture." There are other undercoatings (yellow, blue, and so on) for lower-permeability toroids which are to be used at higher frequen-cies for inductors or conventional transformers.

B) *Magnet test:* All ferrites and powdered-irons are attracted to magnets. Therefore, testing with a permanent magnet is useless.

C) *Electrical test:* Toroidal cores can be tested for permeability, and hence core material, by an inductance measurement. The measure-ment can be made directly with an inductance meter or indirectly by some resonant circuit. Either measurement involves the geometry of the core and the number of turns used in the winding (see Eq 2-1). Rod cores, because of their large air path, and hence reluctance, don't lend themselves as readily to similar measurements. But all of the rods seen by the author at surplus houses and flea markets have been of 125 permeability material and are very likely usable. This material is used in AM radios (loop-stick antennas) and is excellent for transmission line transformers.

D) *Power test:* For those who are not fortunate enough to have access to sophisticated test equipment, another avenue is still available. It is the soak test described in Sec 12.5. It involves applying consider-able power (about 500 W) to a transformer with an unknown core connected back-to-back with a known transformer. One can quickly find out, by the temperature rise, if the ferrite is suitable for use. Ferrites with permeabilities of 40 to 300 exhibit virtually no temperature rise. If the toroids and rods are small (less than 1½-inch in OD or ⅜-inch in diameter for the toroids and rods, respectively), then 200 W would be sufficient. The author has used one of the 12.5:50-Ω designs described in Sec 6.3 for the test.

E) Generally, all of the low-permeability ferrites from the various manufacturers can be interchanged without any significant difference in performance. Therefore, it is a matter of selecting the right permeability range. The following is a list of codes for equivalent ferrites:

1) $\mu = 40$-50: 67, Q2, H54Z
2) $\mu = 125$: 61, 4C4, Q1, H53Z
3) $\mu = 250$-300: 64, 66, K5, H52A, KM1, 250L

Sec 13.3 Winding Rod Transformers

Transmission line transformers with rod cores should find many applications when matching 50-Ω coaxial cable to lower impedances. At these low impedance levels, the coiled transmission lines on rod cores can easily offer sufficient reactance to prevent the unwanted currents and still allow for high-frequency operation. Toroidal cores, with their closed magnetic paths, require fewer turns (and hence yield higher frequency responses), but are not necessary at these low impedance levels. Further, rod transformers are actually easier to wind than their toroidal counterparts.

Fig 13-1 helps in explaining a technique for winding a bifilar transformer such that the winding is tight, and therefore optimized for both its electrical and mechanical properties. First, a single winding, wound as tight as possible, is placed on the rod. A second winding, held in place by soldering it to the first winding, is started on the inside of the first winding. The second winding is then squeezed (or stuffed) between the turns of the first winding. One finds, as the second winding is completed, the two windings are not only tight to each other (resulting in the lowest characteristic impedance possible with the wire), but are literally held fast to the rod. If the windings are somewhat loose on the rod, then small pieces of Scotch no. 27 glass tape, on both ends of the windings, will prevent the rod from falling out. If a trifilar transformer is to be wound, then the first winding should have a spacing between turns of about one wire-diameter. In turn, a quadrifilar transformer should have a spacing of about two wire-diameters, and a quintufilar transformer, about three wire-diameters. In each case, the succeeding windings are soldered, at the ends, with the preceding windings before they are wound on the rods.

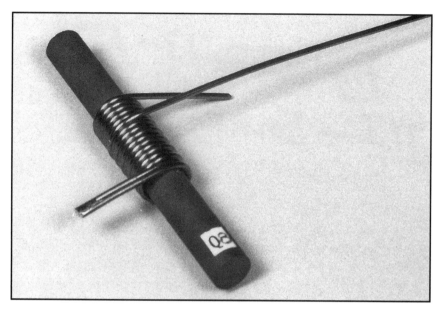

Fig 13-1—The beginning of a tight bifilar winding on a rod core.

Fig 13-2 shows two transformers that are completed and ready for mounting. The top transformer is a 1:4 unun according to Fig 6-1A and (C). Numbers on the wires are a help in making the proper connections. The number 1 terminal is usually the ground connection in the schematics in this book. For the 1:4 unun, the 2,3 terminal is the low-impedance side, and terminal number 4, the high-impedance side. The bottom transformer in Fig 13-2 has a quintufilar winding, resulting in two broadband ratios of 1:1.56 and 1:1.78. The schematic is shown in Fig 10-1C. Winding 9-10 has extra insulation (four layers of Scotch no. 92 tape) in order to improve the high-frequency response of the 1:1.56 ratio. This transformer is designed to match, over a wide bandwidth, 50 Ω to either 32 Ω or 18 Ω. The number 10 terminal is the 50-Ω connection. Although the quintufilar transformer is difficult to wind, there is a regular pattern to the windings, and Fig 7-7B should be of some help in making the connections. Also, tagging the ends of the windings, as shown in the top transformer in Fig 13-2, helps in keeping track of the various windings.

Fig 13-2 also shows another technique which is helpful in keeping the windings in place. All of the wires, grasped by a pair of pliers, are unwound about a quarter turn. They are then bound together with a small

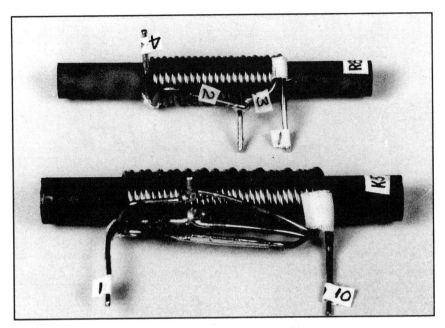

Fig 13-2—Two rod transformers ready for mounting. The top transformer is a 1:4 unun matching 12.5 Ω to 50 Ω. The bottom transformer is an unun with two broadband ratios, matching 32 Ω to 50 Ω or 18 Ω to 50 Ω.

section of Scotch no. 27 glass tape and rolled back onto the rod. This tape helps keep the windings in place when making the proper connections. The pliers will be found to be helpful since copper wire becomes work-hardened in handling and takes considerable force in the unwinding and rewinding process.

And finally, a few words are given on tapping windings. In some cases, impedance ratios are required which involve tapping the windings. For example, a trifilar winding connected as in Fig 7-14 yields a ratio of 1:2.25. Tapping winding 5-6, close to terminal 6, can yield a broadband ratio very near 1:2. Successful taps have been made by first filing about a ⅛ inch-wide groove around the wire with the edge of a small, fine file. Then a copper strip or a 14- or 16-gauge wire, which has one end flattened about ⅜-inch long, is wrapped around the groove and soldered. The soldered connection is then rendered smooth by the edge of the file. Two or three sections of Scotch no. 92 tape are then placed over the soldered area in order to provide mechanical and electrical protection for the adjacent windings. Fig 13-3 shows a bifilar

Fig 13-3—A tapped, bifilar transformer yielding a ratio of 1 :3.

transformer which is tapped to yield a ratio of 1:3. The chassis connector is on the low-impedance side of the transformer.

Sec 13.4   Winding Toroidal Transformers

Toroidal cores offer the greatest margins in bandwidth and power because of their closed magnetic paths. This allows for the use of higher permeabilities (which rods do not allow), and hence much shorter windings then their rod counterparts. With ML or H Imideze wire and/or coaxial cable, toroidal transformers are considered the "top-of-the-line." Since so little flux enters the core of a transmission line transformer, the objectives with toroidal transformers are to use the smallest cores and the highest permeability allowed by the size of the conductors and the requirement on efficiency (see Secs 2.2 and 11.3). In many applications, cores of only 1½-inch in OD can handle the full legal limit of power allowed for Amateur Radio use. Larger cores have to be used at the higher impedance levels in order to obtain the required choke inductance and characteristic impedance. There are two major differences in winding toroidal transformers, as compared to rod transformers, and they are:

1) The conductors (except for a single coaxial cable) are bound together and wound as a ribbon.

2) Because more than one conductor is wound at a time, and considerably more bending and unbending is experienced, work-hard-

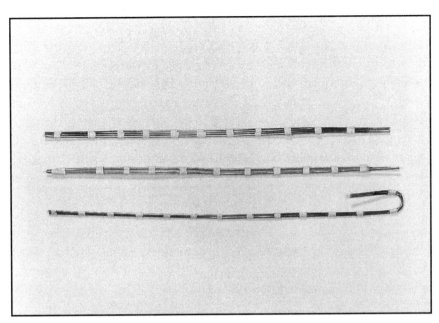

Fig 13-4—Three typical ribbons used in toroidal transformers. The bottom bifilar ribbon is shown to suggest a beginning in the winding process (see text).

ening of the wire is much more severe. Thus, considerably more force is required in the winding process. The thumb and pliers become indispensable tools.

Fig 13-4 shows three typical ribbons used in toroidal transformers. The endings (on the right) have the insulation scraped off to about ¾ inch. The wires are held in place with Scotch no. 27 glass tape every ¾ inch. The quadrifilar ribbon (on the top) has a no. 14 wire on the second one down from the top. The other three are no. 16 wires. This ribbon was optimized for a 1:1.78 toroidal transformer matching 28 Ω to 50 Ω (see Fig 7-28B). The ribbon in the middle of Fig 13-4 is composed of two no. 14 wires, each covered with about a 20-mil wall Teflon tubing. The insulation is primarily used to separate the wires in order to obtain the desired characteristic impedance of 100 Ω (see Fig 5-2). The ribbon at the bottom is composed of no. 14 wires, held close together by the glass tape clamps, resulting in a characteristic impedance of 40 Ω. If one of the wires were covered with two layers of Scotch no. 92 tape (6 mils of insulation—see Fig 5-2), the characteristic impedance would approach 50 Ω. This ribbon is primarily shown to suggest a beginning in the process of winding a toroidal transformer. The ribbon is first placed on the outside diameter (perpendicular to it) of the toroid with about 1½ inches overhang. The larger part of the wire is then bent downward 90 degrees. The ribbon is then placed on the inner diameter and bent another 90 degrees. This is the stage shown in Fig 13-4. Then the ribbon is placed back on the outer diameter. The large end is then fed back (with some bending) through the inside of the core to complete the first turn. The thumb and a pair of pliers help in making the first turn tight to the core. After two or three turns, the winding is well anchored to the core and then can be completed quite easily.

Fig 13-5 is a photograph of two toroidal transformers showing their various connections. The one on the left has six trifilar turns on a 1½-inch OD, no. 64 toroid. The schematic is shown in Fig 7-20. Windings 1-2 and 5-6 are no. 16 wire. Winding 3-4 is no. 14 wire and is tapped at 1, 3 and 5 turns from terminal 3. This transformer has broadband ratios of 1:1.2, 1:1.56, 1:2 and 1:2.25. The tags help to identify the terminals. The 2,5 connection is untagged. The transformer on the right in Fig 13-5 has five quintufilar turns, also on a 1½-inch OD, no. 64 toroid. The schematic is shown in Fig 7-8A. Winding 9-10 is no. 14 wire and the other four are no. 16 wire. An input connection is also made to terminal 7, resulting in two broadband ratios of 1:1.56 (input to terminal 9) and 1:2.78 (input to terminal 7). This transformer is designed to match 50 Ω to either 75 Ω or 140 Ω. The various tags help in identifying the connections.

Fig 13-5—Two toroidal transformers, displaying the various connections. The 1:2.25 transformer on the left uses the schematic in Fig 7-20. The 1:1.56 transformer on the right uses the schematic in Fig 7-8A.

Sec 13.5   Constructing Low-Impedance Coaxial Cable

When designing transmission line transformers that require characteristic impedances less than 25 Ω, stripline or low-impedance coaxial cable are usually resorted to. Neither of these transmission lines are readily available commercially in the range of 5 Ω to 35 Ω. If one has access to a machine shop, then a sheet of copper can be slit to the widths required for the specific characteristic impedances (see Fig 4-3). Characteristic impedances as low as 5 Ω have been obtained (by the author) using ⅜-inch strips of copper and one layer of Scotch no. 92 tape for insulation. With homemade, low-impedance coaxial cable, values from 9 Ω to 35 Ω have been easily constructed from readily available components. With further effort, values as low as 5 Ω should be achievable.

Low-impedance coaxial cables have a decided advantage over tightly-wound wire transmission lines. This is because: 1) the currents can be considerably larger, since they are evenly distributed about the inner conductor and the outer braid, and 2) the voltage breakdowns are considerably larger (rivaling that of RG-8/U), since several layers of Scotch no. 92 tape are generally used in achieving the desired characteristic impedances. Table 4-1 shows the results (obtained by the author) on the characteristic impedance as a function of various combinations of wire size and insulation thickness. Although not shown in the table, a single layer of Scotch no. 92 tape on a no. 10 inner conductor was found to produce a characteristic impedance of 9 Ω. ML or H Imideze wire, without any extra insulation, should yield even lower impedances. The outer braids of all the cables in Table 4-1 were tightly wrapped with Scotch no. 92 tape. Without this outer wrap, the characteristic impedance was found to be greater by about 25%.

An indispensable tool for constructing these cables is shown in Fig 13-6. This U-frame, mounted in a vise, is constructed out of ¼-inch plywood and has a usable span from 15 inches to 30 inches. Other thicknesses of plywood, obviously, can be used. The vertical struts are 9 inches in height. The horizontal piece is 24 inches long. Three different settings are available on the right side, resulting in 2-inch changes in the span. A third hole, which is made at the bottom of the struts, allows them to be tilted outward at about 45 degrees. This allows the span to be increased to 30 inches. The inner conductor wire is held

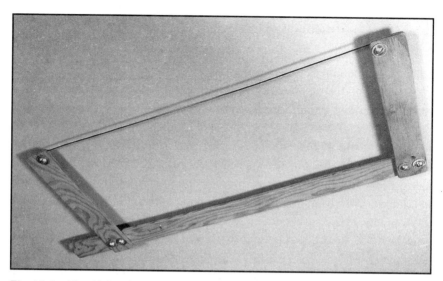

Fig 13-6—The U-frame used in constructing low-impedance coaxial cable.

taut by the two ¼-20 bolts at the top of the struts. The insulation on the inner conductor can be put on by one of the following two methods:

1) *Longitudinal:* This is like rolling up a carpet or a window shade. Scotch no. 92 or no. 27 is attached at one edge, along the length of the wire. The ½-inch tapes, which are then rolled on, put about two layers on 12- to 16-gauge wire. If four layers are required, then this is done twice. As a result, the no. 16 wire will have a little more than four layers on it, and the no. 12 wire, a little less. Since the characteristic impedance is a function of the log of the ratio of the diameter of the outer braid to the diameter of the inner conductor (and therefore not sensitive to small differences in actual diameters), the desired result of two or four layers is practically realized. This is the method used by the author.

2) *Spiral:* This is like taping a baseball bat. By carefully controlling the pitch of the spiral (and hence the overlap), one is able to place, quite accurately, the desired number of layers. For four layers, it is recommended that this process be done in two steps. This method is also easy to use because of the tall struts in the U-frame.

After the insulation has been placed on the inner conductor, it is removed from the U-frame and made ready for the outer braid. Except for the 30- to 35-Ω cable in Table 4-1, which use the outer braid from

RG-58/U, all of the others use the braids from smaller cables such as RG-122/U. Flat ⅛-inch braid, which is opened up easily by the point of a pencil, has also been used successfully. The only requirement on the outer braid is that it has practically 100% coverage. That is, the inner conductor should not be seen when the cable is wound around a core. Small end-caps of ⅛-inch copper strip or 14- or 16-gauge wire provide adequate contacts to the braid. If the braid is to be tightly wrapped, the whole structure is placed back on the U-frame and a spiral winding of Scotch no. 92 tape (or practically any other tape) is applied. Tapping of the inner conductor is made possible by using two sections of outer braids and soldering them together around the tap (even the center conductor of a coaxial cable can have a longitudinal potential gradient!).

Sec 13.6   The Care and Handling of Ferrite Transformers

Nickel-zinc ferrites (which are the ones to be used in power transmission line transformers) have bulk resistivities in the range of 10^5 to 10^9 Ω-cm. This means that they are excellent insulators and do not require extra insulation on them from an electrical standpoint. Even bare wire (as long as it doesn't touch its neighboring turns) can be used without any extra precautions. Coatings on the cores are mainly used for the mechanical protection of the wires. With automatic winding machines especially, the insulation can be harmed by rough surfaces and sharp edges, resulting in the possibility of a short circuit. With the smooth surfaces and well-chamfered edges of modern toroids, the harm, during winding, is considerably reduced or even eliminated (especially when done by hand).

Another misconception regarding ferrite cores or beads is the need for extra protection from the environment. Since ferrite is a ceramic, there is practically no moisture adsorption on the surface or penetration of it below the surface. Even if there were, the properties of the ferrite would be unchanged. For years, the author has used ferrite transformers mounted in miniboxes with their covers in place without using the self-tapping screws or tape along the exposed edges. These transformers have been subjected to all kinds of weather without any noticeable changes in performance. The only precaution is to keep the rain and snow off the transformers. The coiled transmission lines would not like this.

And finally, what happens to ferrites (which are brittle) when they are broken into several large pieces? If the pieces are large enough (and not too many) they can be glued together and the core will perform as well as before. The precaution here is to glue the parts as tight as possible in order to eliminate the high reluctance of a sizable gap (air space) between the ferrite pieces.

Chapter 14

Summary Statements

T his chapter is a digest of important investigative results observed during experimentation by the author. Significant points concerning the capabilities of ferrite-core transformers are also reviewed. Some of the comments are contrary to conventional wisdom on the subject. It is hoped that the information presented in this book will assist the circuit designer and the analyst, and stimulate additional research in the field.

Comment 1

Some of the more important additions in this second edition are: (A) an in-depth study of Guanella's approach to transmission line transformers, (B) a comparison between the approaches of Guanella and Ruthroff, (C) a new look at balun transformers, (D) parallel transformers, (E) broadband, dual-ratio transformers, (F) fractional-ratio balun transformers and (G) many more practical transformer designs.

Comment 2

Since the publication of the first edition, analytical and experimental results (by the author) on a great variety of transformers revealed some interesting comparisons of the approaches by Guanella (ref 1)[1] and Ruthroff (ref 9) to transmission line transformer theory and applications. Their papers are considered the cornerstones of this class of transformers. Guanella was concerned with the matching of the push-pull output of a 100-W vacuum-tube amplifier to coaxial cable in the 100- to 200-MHz range. Ruthroff's research, on the other hand, was mainly concerned with broadband matching at low signal levels to coaxial cable, radio, millimeter waveguide and early optical fibers (for the emerging digital techniques). Many of his transformers performed into the gigahertz range and handled nanosecond pulses. Besides the

---

[1]Each reference in this chapter can be found in Chapter 15.

power differences in their transformers, there are other important ones (noted throughout this book), which are briefly listed as follows:

A) Guanella's 1:4 transformer, which is actually a bilateral balun, is composed of two transmission lines in a parallel-series arrangement. When the transmission lines are effectively terminated in their characteristic impedances, the high-frequency response is mainly limited by the parasitics, since in-phase voltages are summed at the high-impedance side (Sec 9.3.3).

B) Ruthroff has two versions of the 1:4 transformer. One is a balun (the high-impedance side always balanced) and depends upon a negative potential gradient along the transmission line. The other is an unun (unbalanced-to-unbalanced transformer) and depends upon a positive potential gradient. Both transformers sum a direct voltage with a delayed voltage and, as such, have a built-in, high-frequency cut-off, regardless of the value of the characteristic impedance or the parasitics (Sec 1.4).

C) Guanella's 1:1 balun uses a single transmission line. Ruthroff's 1:1 balun, which also uses a single transmission line, has a third wire connected to ground to (again) present a negative potential gradient. On a rod, this arrangement appears as a trifilar winding. Further, this third wire can present some problems if the choking action of the balun becomes insufficient to prevent conventional transformer currents (Sec 9.2).

D) Guanella's method of connecting transmission lines in parallel-series arrangements also allows for very broad bandwidths with large impedance ratios and at high impedance levels. The ratios are restricted to $1:n^2$, where n, the number of transmission lines, is an integral number. High impedance levels, which require more choking reactance (and hence more turns) for isolation, are not as much of a problem here, because in-phase voltages are summed. Since they are in phase, the high-frequency response is not as dependent upon the length of the transmission line as it is with the Ruthroff 1:4 transformers.

Comment 3

The multiwinding ununs (like trifilar, quadrifilar, and so on) proposed by the author are really an extension of Ruthroff's unun, since they rely on "boot-strap" connections (Sec 1.2) and the resulting positive potential gradients (Sec 7.2). Depending upon the number of windings, how they are connected, whether they are tapped and the impedance level, broad bandwidth ratios from 1:1 to 1:16 (and any in between) are readily available. These multiwinding ununs, in series with baluns, allow for a variety of impedance ratios for baluns other than $1:n^2$ (where n is an integral number).

Comment 4

Multiwinding ununs, with ratios less than 1:4, exhibit very broad bandwidths (Fig 7-9). This is because the windings are in series-aiding at the low-frequency end and, as such, allow for shorter transmission lines. Further, the delayed voltages which traverse the transmission lines are a smaller component of the output voltages. As the number of windings increases, the impedance ratio decreases, and the resulting bandwidth increases.

Comment 5

Transmission line transformers are basically low-impedance devices. Their high-frequency responses are dependent upon optimizing the characteristic impedance of the transmission lines and minimizing the parasitics and the length of the transmission lines (particularly with Ruthroff designs). In practice, characteristic impedances as low as 5 Ω with stripline (Fig 4.3) and as high as 200 Ω with spaced windings (Fig 5.2) are readily obtainable. For the Ruthroff 1:4 transformers, this results in broad bandwidths in the impedance level range of 2.5:10-Ω to 100:400-Ω (Sec 1.4). Guanella's modular technique of employing parallel-series connected transmission lines allows for higher impedance ratios, thus extending the impedance level by a factor of two to three. For example, with five transmission lines, broadband operation is possible up to 1000 Ω. Further, since his transformers sum in-phase voltages, the bandwidths of his transformers are considerably greater than the 1:4 transformers of Ruthroff (Sec 1.3).

Comment 6

Guanella's transformers, which are basically baluns, can be converted to broadband hybrids and ununs by accounting for the effect on the low-frequency response when two ground connections are made on the transformers. This is particularly important with his 1:4 balun when the two transmission lines are wound on a single core (Sec 2.3). Without the use of two separate cores or a series 1:1 balun, the low-frequency response as an unun is seriously impaired. His 1:9 and 1:16 baluns are not affected as much, since they always employ separate cores for each transmission line. Even then, these high-ratio transformers can benefit some by the use of a series 1:1 balun for added isolation. Further, without the series 1:1 balun, the bottom transmission lines on these ununs (Figs 6-2 and 8-1) do not require a magnetic core, since no potential difference exists between their input and output terminals. This is also true when his 1:4 balun, using separate cores, is connected as an unun (Fig 6-2).

Comment 7

Loss measurements on Ruthroff 1:4 ununs show that low-permeability, nickel-zinc ferrite cores give the highest efficiencies. Because of the flux-canceling effect of the transmission line currents, efficiencies exceeding 98% over nearly two decades in frequency are possible by using proper ferrites. Even 99% efficiencies can be realized over a large portion of the passbands (Sec 11.3). Guanella transformers, which do not rely on the longitudinal voltage gradients of the Ruthroff transformers (but have various gradients depending on where the ground connections are made), should exhibit the same high efficiencies.

Comment 8

Analyses and conventional wisdom have assumed that the characteristics of the core and the number of bifilar turns of Ruthroff's 1:4 transformers determine only the low-frequency responses. A second belief was that the characteristic impedance and length of the transmission line determine only the high-frequency responses. However, accurate measurements show that the core and the transmission line mode are effective over the entire passband of these transformers (Sec 3.4).

Comment 9

Earlier investigators have advocated that transmission lines be twisted to reduce the characteristic impedance for operation at low impedance levels. Experimental data by the author is in contradiction of this belief. In power applications, where thick-wire transmission lines (no. 12 to no. 16 wire) are used, best operation is realized by placing the transmission lines as close as possible to one another (Sec 3.3).

Comment 10

In general, any type of winding placed on a rod or toroid can give constant impedance transformation ratios over two or three of the Amateur Radio bands. By using windings with optimal characteristic impedances, the bandwidths can be more than doubled, so that at least six bands can be covered (Sec 3.2). With Guanella transformers that have very little parasitics, the difference in bandwidth is even more dramatic.

Comment 11

In low-power or small-signal applications, efficiency is generally not a consideration. Here, high-permeability ferrite toroids, which require fewer turns for good low-frequency response, can be used advantageously. Bandwidths in this class of operation can greatly exceed that of high-power transmission line transformers. An interesting nickel-zinc ferrite to explore is CMD 5005, produced by Magnetic Ceramics (Secs 11.3 and 11.5). It has a relatively high permeability of 1400, a high bulk resistivity of 7×10^9 and exhibits reasonably high efficiency.

Comment 12

From many on-the-air discussions, it is apparent that almost everyone still perceives transmission line transformers as conventional transformers. As such, they express concern about the properties of cores such as loss, size, saturation, frequency response and IMD (intermodulation distortion). The term magnetizing current is also brought into the picture. But all of these terms are only related to the low-frequency models of these devices, where virtually no energy is transmitted by a transmission line mode. In other words, the reactances of the coiled or ferrite-beaded transmission lines are much less than that of the load they see. In properly designed transformers, these reactances are much greater than the loads the transmission lines see, thus isolating the output from the input and allowing for the efficient transmission line mode. In response to the expressed concerns regarding these devices, the following statements are offered. When using properly designed transmission line transformers:

A) The size of the core is mainly determined by the size and spacing of the conductors.

B) Ferrite-core transformers are superior to air-core transformers (Sec 9.2.3).

C) There is virtually no flux in the core, so saturation and IMD are irrelevant considerations.

D) The high-frequency response of the transformer is independent of the high-frequency response of the core (Sec 3.5).

E) The bifilar Guanella 1:1 balun is the preferred 1:1 balun (Sec 9.2.2).

F) Low-permeability ferrites are necessary for very high efficiencies (Sec 11.3).

G) The transmission line transformer is basically an RF choke and a configuration of transmission lines.

Comment 13

The design chapters in this second edition (Chapters 6 through 10) not only present a large variety of different transformer designs, but many designs apparently performing the same function. The problem then, for many, is one of selecting the right design for the application at hand. Here are some broad guidelines for making that choice.

A) Baluns

1) Guanella 1:1 baluns have bifilar windings and are easier to design and construct than Ruthroff 1:1 baluns, which have three windings. With sufficient reactance from the coiled transmission lines, the Guanella transformer can meet any of the requirements of a 1:1 balun (Sec 9.2). Further, it does not experience the high core flux possible at very low frequencies with the Ruthroff 1:1 balun.

2) Toroidal 1:1 baluns, because of their closed magnetic paths and ability to effectively use higher-permeability ferrites (250 to 300), allow for much greater margins in bandwidth and power levels (Sec 9.2.1). They are the "top-of-the-line." But rod 1:1 baluns with sufficient bifilar turns (like 12 to 14 on a ½-inch diameter rod, 3 to 4 inches long, $\mu = 125$) can practically meet all of the objectives in the 1.5-MHz to 30-MHz range (Sec 9.2.1). Increasing the permeability of the ferrite rod has negligible effect on its performance (Sec 2.5).

3) Guanella baluns (other than the 1:1 balun) achieve wider bandwidths and higher impedance ratios and can operate at higher impedance levels than the Ruthroff balun. Specifically, the Guanella 50:200-Ω balun has a very much higher frequency response than the popular Ruthroff design (Sec 9.3.1). When the balanced load is grounded at its midpoint, the Guanella balun loses some of its low-frequency capability. This is particularly true of the 1:4 Guanella balun which uses a single core. Separate cores or a series 1:1 balun are needed in order to restore the low-frequency response. On the other hand, the Ruthroff 1:4 balun maintains its same low-frequency response and

increases greatly its high-frequency response. It now appears to act like a Guanella balun (Sec 9.3).

4) Ratios other than 1:n^2 (where n is an integral number) can easily be achieved by connecting various low-ratio ununs in series with Ruthroff or Guanella baluns (Sec 9.5).

B) Ununs (unbalanced-to-unbalanced transformers)

1) Because of the technique of adding in-phase voltages, Guanella baluns can be converted to ununs (by adding extra isolation) and still achieve wider bandwidths, higher impedance ratios, and operate at higher impedance levels than Ruthroff ununs.

2) The 12.5:50-Ω (1:4) unun: If the bandwidth of interest is from 1.5 MHz to 30 MHz, then the Ruthroff design, which employs a single transmission line and therefore has the advantage of simplicity, is recommended (Sec 6.3). A rod transformer, especially using ML or H Imideze wire for added voltage-breakdown margin, can perform very well at this impedance level and frequency range. A toroid with a few less turns of 20-Ω coaxial cable (Sec 4.3) on a 1½- to 1¾-inch OD, toroid (μ = 250 to 300) would have a greater bandwidth and higher power capability. For much wider bandwidths, the Guanella balun (with added isolation) is again recommended. For the VHF and UHF bands, straight, beaded coax should be investigated.

3) Impedance ratios less than 1:4: The designs which are shown in Chapter 7 are really an extension of Ruthroff's 1:4 unun. The higher-order windings (trifilar, quadrifilar, quintufilar, and so on) do very well on rod cores in the range of 1.5 MHz to 30 MHz. With toroidal cores, the bandwidths can, again, be greater.

C) Other design considerations:

Much of the design information and most of the examples in this second edition have been concerned with power transmission line transformers operating in the 1.5- to 30-MHz range. Many of the examples (especially the Guanella transformers) have exceeded the upper frequency limit by a factor of at least two or three times. For those interested in transformers for the VHF and UHF bands, or efficient transformers operating at impedance levels in the 200- to 1000-Ω range, the following recommendations are offered:

1) Efficiency can be traded off for low-frequency response when working at high impedance levels. Transformers using ferrites with

permeabilities of 40 to 50 still maintain their very high efficiencies at impedance levels at least up to 200 Ω (Sec 11.3). It is safe to assume that efficiencies with these ferrites still remain high at the 1000-Ω level. With permeabilities of 40 to 50, the low-frequency response suffers by a factor of about six when compared to that of a ferrite with a permeability of 290 (which has more loss at higher impedance levels). The result is that the low-frequency response is changed from 1.5 MHz to 9 MHz for higher efficiency.

2) Guanella's approach has the greatest chance of providing wideband, power transmission line transformers in the VHF and UHF bands since it adds in-phase voltages. With their coiled transmission lines effectively terminated in their characteristic impedances, the high-frequency response is predominantly determined by the parasitics. Two important ways of reducing the parasitics are by increasing the spacing between adjacent bifilar turns or by resorting to straight, ferrite-beaded coaxial cable.

Comment 14

While much new information appears in this revised edition, further important work can still be carried out in this field. Areas needing further study are:

A) A new look at ferrites when used in transmission line transformers. Practically all of the present developments, characterizations and specifications are only related to their use in conventional transformers or inductors.

B) Research and development on new ferrites for use in power transmission line transformers. Ferrites with permeabilities in excess of 300 are of particular interest.

C) Improved analytical models involving low-impedance transmission lines of stripline and coaxial cable. Measurements show that the optimum characteristic impedances are only 80% to 90% of predicted values.

D) Investigations on other models for obtaining fractional ratios. These could involve transmission line transformers operating in a series mode (ref 17) instead of the parallel mode proposed by the author (Chapter 7).

E) Further investigations on power ratings and reliability. These will especially involve ferrites with permeabilities in excess of 300.

Chapter 15

References

[1]Guanella, G., "Novel Matching Systems for High Frequencies," *Brown-Boveri Review*, Vol 31, Sep 1944, pp 327-329.

[2]Fubini, E. G. and P. J. Sutro, "A Wide-Band Transformer from an Unbalanced to a Balanced Line," *Proc IRE*, Vol 35, Oct 1947, pp 1153-1155.

[3]Rudenberg, H. Gunther, "The Distributed Transformer," Res Div, Raytheon Manuf Co, Waltham, MA, Apr 1952.

[4]Rochelle, Robert W., *The Review of Scientific Instruments 23*, 298 (1952).

[5]Lewis, I. A. D., *The Review of Scientific Instruments 23*, 769 (1952).

[6]Brennan, Alan T., "A UHF Balun," RCA Laboratories Div, Industry Service Laboratory, LB-911, May 5, 1953.

[7]Talkin, A. I. and J. V. Cuneo, "Wide-Band Transformer," *The Review of Scientific Instruments*, Vol 28, No. 10, 808, Oct 1957.

[8]Roberts, W. K., "A New Wide-Band Balun," *Proc IRE*, Vol 45, Dec 1957, pp 1628-1631.

[9]Ruthroff, C. L., "Some Broad-Band Transformers," *Proc IRE*, Vol 47, Aug 1959, pp 1337-1342.

[10]Turrin, R. H., "Broad-Band Balun Transformers," *QST*, Aug 1964, pp 33-35.

[11]Matick, R. E., "Transmission Line Pulse Transformers—Theory and Applications," *Proc IEEE*, Vol 56, No. 1, Jan 1968, pp 47-62.

[12]Pitzalis, O. and T. P. Couse, "Practical Design Information for Broadband Transmission Line Transformers," *Proc IEEE*, Apr 1968, pp 738-739.

[13]Pitzalis O. and T. P. Couse, "Broadband Transformer Design for RF Power Amplifiers," *US Army Tech Rept ECOM-2989*, Jul 1968.

[14]Turrin, R. H., "Applications of Broad-Band Balun Transformers," *QST*, Apr 1969, pp 42-43.

[15]Pitzalis, O., R. E. Horn and R. J. Baranello, "Broadband 60-W Linear Amplifiers," *IEEE Journal of Solid State Circuits*, Vol SC-6, No. 3, Jun 1971, pp 93-103.

[16]Krauss, H. L. and C. W. Allen, "Designing Toroidal Transformers to Optimize Wideband Performance," *Electronics*, Aug 16, 1973.

[17]London, S. E. and S. V. Tomeshevich, "Line Transformers with Fractional Transformation Factor," *Telecommunications and Radio Engineering*, Vol 28/29, Apr 1974.

[18]Granberg, H. O., "Broadband Transformers and Power Combining Techniques for RF," *Motorola Application Note AN-749*, 1975.

[19]Sevick, J., "Simple Broadband Matching Networks," *QST*, Jan 1976, p 20.

[20]Sevick, J., "Broadband Matching Transformers Can Handle Many Kilowatts," *Electronics*, Nov 25, 1976, pp 123-128.

[21]Blocker, W., "The Behavior of the Wideband Transmission Line Transformer for Nonoptimum Line Impedance," *Proc IEEE*, Vol 65, 1978, pp 518-519.

[22]Sevick, J., "Transmission Line Transformers in Low Impedance Applications," *MIDCON 78*, Dec 1978.

[23]Dutta Roy, S. C., "Low-Frequency Wide-Band Impedance Matching by Exponential Transmission Line," *Proc IEEE* (Letter), Vol 67, Aug 1979, pp 1162-1163.

[24]Dutta Roy, S. C., "Optimum Design of an Exponential Line Transformer for Wide-Band Matching at Low Frequencies," *Proc IEEE* (Letter), Vol 67, No. 11, Nov 1979, pp 1563-1564.

[25]Irish, R. T., "Method of Bandwidth Extension for the Ruthroff Transformer," *Electronic Letters*, Vol 15, Nov 22, 1979, pp 790-791.

[26]Kunieda, H. and M. Onoda, "Equivalent Representation of Multiwire Transmission-Line Transformers and its Applications to the Design of Hybrid Networks," *IEEE Trans on Circuits and Systems*, Vol CAS-27, No. 3, Mar 1980, pp 207-213.

[27]Granberg, H. O., "Broadband Transformers," *Electronic Design*, Jul 19, 1980, pp 181-187.

[28]Collins, R. E., *Foundations for Microwave Engineering*, New York: McGraw Hill, 1966, Chap 5.

[29]*The ARRL 1990 Handbook for the Radio Amateur*, Newington: ARRL, 1989, pp 11-13 to 11-15.

[30]Takei, T., "Review of Ferrite Memory Materials in Japan," Ferrites, *Proc of the Intl Conf*, ed. Y. Hoshimo, S. Jida, and M. Sugimoto, Baltimore: University Park Press, pp 436-437.

[31]Snoek, J. L., *New Developments in Ferromagnetic Materials*, New York: Elsevier, 1947.

[32]Stone, Jr, H. A., "Ferrite Core Inductors," *Bell System Tech Jour*, Vol 32, Mar 1953, pp 265-291.

[33]Slick, P. I., US Pat No. 3,533,949; filed Nov 21, 1967, issued Oct 13, 1970.

[34]Geldart, W. J., G. D. Haynie, and R. G. Schleich, "A 50-Hz – 250-Mhz Computer Operated Transmission Measuring Set," *Bell Systems Tech Jour*, Vol 48, No. 5, May/Jun 1969.

[35]Geldart, W. J. and G. W. Pentico, "Accuracy Verification and Intercomparison of Computer-Operated Transmission Measuring Sets," *IEEE Trans on Instr and Meas*, Vol IM-21, No. 4, Nov 1972, pp 528-532.

[36]Geldart, W. J., "Improved Impedance Measuring Accuracy with Computer-Operated Transmission Measuring Sets," *IEEE Trans on Instr and Meas*, Vol IM-24, No. 4, Dec 1975, pp 327-331.

About the Author

Jerry Sevick's favorite subject is transmission line transformers. Many of his developments that are featured in this book were made in his home laboratory. He is also a licensed Amateur Radio operator and holds the call W2FMI. Jerry has authored over a dozen magazine articles on the subject of antennas. He is noted for a classic series on short vertical antennas that appeared in *QST*, the Amateur Radio journal of the American Radio Relay League.

Jerry earned a BS in education from Wayne State University in Michigan and a PhD in applied physics from Harvard University. Jerry's career reveals an interest in a variety of fields: He taught physics at Wayne State University, and spent two years as a weatherman at a network television station in Detroit. In 1956, Jerry joined the staff at AT&T Bell Laboratories in Murray Hill, New Jersey. He was a supervisor in groups working on high-frequency transistor and integrated-circuit development, and later he served as Director of Technical Relations, acting as liaison between the technical staff and outside organizations.

Having retired from Bell Labs in 1984, Jerry remains active in many areas. While he was chairman of the Transnational Relations Committee of the IEEE, he arranged the first technical exchanges with the People's Republic of China in 1977. He is a Technical Advisor for the American Radio Relay League and is a member of IEEE, Sigma Xi, Sigma Pi Sigma and Phi Delta Kappa.